Local Realities and Environmental Changes in the History of East Asia

Environmental history has evolved into a well-established historical subfield which has broadened the horizons of historical research, beyond human affairs, to include the study of human interactions with natural and man-made environments. This broadened scope has attracted scholars from many different fields; a development which is reflected by this volume as it highlights the recent studies on East Asian environmental history by scholars of history, economic history, political ecology, sociology and environmental studies.

This book examines the local realities and environmental changes in East Asia, and is one of a few publications in English on the subject. Contributors apply rich historical material, maps and statistical data to reveal the local environmental realities infused by global perspectives. Part I deals with attitude toward nature, focusing on the soundscape conceived by traditional Chinese literati and on "industrious revolution" in Tokugawa Japan. Part II includes four case studies which respectively discuss the hydraulic management and political ecology in the Yongle reign (1403–1424), the "Woosung Bar" controversy in the 1870s, the expansion of Daihaizi Reservoir in Xinjiang in the 1950s, and interactions between the indigenous communities and NGOs in Hualien, Taiwan. Part III presents case studies of Japan dealing with natural disasters: volcano eruption, floods and the human actions around Tokyo since the eighteenth century. These chapters and the insights they offer provide the reader with the most recent research on East Asian environmental history.

Covering the geographical areas of Japan, North and Northwest China, the Lower Yangzi Delta and Taiwan, and the timeframe spanning the seventh century BC to the present day, the book will be of great interest to anyone studying the history of East Asia, environmental history or environmental studies.

Ts'ui-jung Liu is an Adjunct Research Fellow at Academia Sinica, Taiwan.

中央研究院
ACADEMIA SINICA

Academia Sinica on East Asia
Published in association with Academia Sinica, Taiwan
Series Editor: Dr. Fan-Sen Wang
Vice-President, Academia Sinica, Taiwan

Academia Sinica was founded in 1928 in Nanjing, China. Over the next decade, ten institutes were founded, including three institutes in the humanities and social sciences. Researchers continued their work through the War of Resistance against Japan (1936–1945) and the civil war (1945–1949), and in 1948 Academia Sinica followed the government of the Republic of China to Taiwan. Only two of its institutes, the Institute of History and Philology and the Institute of Mathematics, were initially reestablished in Taiwan. However, over the following years Academia Sinica developed into a world-class research institution with 31 institutes and research centers divided among three divisions: Mathematics and Physical Sciences, Life Sciences, and Humanities and Social Sciences.

The Division of Humanities and Social Sciences now consists of 11 institutes and a research center. Its scholars conduct research in the areas of archeology, history, literature, linguistics, philology, philosophy, anthropology, economics, sociology, political science, and law, as well as conducting interdisciplinary research.

The publication series *Academia Sinica on East Asia* features monographs by scholars in the humanities and social sciences at Academia Sinica.

Taiwan in Japan's Empire-Building
An institutional approach to colonial engineering
Hui-yu Caroline Ts'ai

Divine Justice
Religion and the development of Chinese legal culture
Paul R. Katz

The Poet-historian Qian Qianyi
Lawrence C. H. Yim

Beyond Confucian China
The rival discourses of Kang Youwei and Zhang Binglin
Young-tsu Wong

Dandyism and Transcultural Modernity
The Dandy, the Flaneur, and the Translator in 1930s Shanghai, Tokyo, and Paris
Hsiao-yen Peng

The Economy of Lower Yangzi Delta in Late Imperial China
Connecting money, markets, and institutions
Billy K. L. So

Environmental History in East Asia
Interdisciplinary perspectives
Edited by Ts'ui-jung Liu

From Eileen Chang to Ang Lee
Lust/caution
Edited by Hsiao-yen Peng and Whitney Crothers Dilley

Local Realities and Environmental Changes in the History of East Asia
Edited by Ts'ui-jung Liu

Local Realities and Environmental Changes in the History of East Asia

Edited by Ts'ui-jung Liu

Routledge
Taylor & Francis Group
LONDON AND NEW YORK

First published 2016
by Routledge
2 Park Square, Milton Park, Abingdon, Oxon OX14 4RN

and by Routledge
711 Third Avenue, New York, NY 10017

First issued in paperback 2017

Routledge is an imprint of the Taylor & Francis Group, an informa business

British Library Cataloguing in Publication Data
A catalogue record for this book is available from the British Library

Library of Congress Cataloging in Publication Data
Local realities and environmental changes in the history of East Asia / edited by Ts'ui-jung Liu.
pages cm. – (Academia Sinica on East Asia)
Includes bibliographical references and index.
1. Human ecology–East Asia–History. 2. Human ecology–East Asia–History–Case studies. I. Liu, Cuirong, editor of compilation.
GE185.E18L64 2016
304.2095–dc23 2015010394

ISBN 13: 978-1-138-48254-8 (pbk)
ISBN 13: 978-1-138-90615-0 (hbk)

Typeset in Times New Roman
by Wearset Ltd, Boldon, Tyne and Wear

Contents

Figures

Tables

Contributors

Beatrice M. Bodart-Bailey is Professor Emeritus, Otsuma Women's University, Tokyo and Visiting Research Fellow, School of Culture, History and Language, ANU College of Asia and the Pacific, Australian National University, Canberra.

Chun-Chieh Chi 紀駿傑 is Professor at the Department of Ethnic Relations and Culture, National Dong-Hwa University, Taiwan.

Cho-ying Li 李卓穎 is Associate Professor at the Institute of History, National Tsing Hua University, Taiwan.

Ts'ui-jung Liu 劉翠溶 is an Academician of Academia Sinica (elected in 1996). She retired in January 2014 from a position of Distinguished Research Fellow and is currently an Adjunct Research Fellow at the Institute of Taiwan History, Academia Sinica, Taiwan.

Satoshi Murayama 村山聡 is Professor of Economic and Environmental History, Kagawa University, Japan.

M. William Steele is Professor of History at International Christian University, Japan.

Lihua Wang 王利華 is Professor at the Center of Environmental History of China, Nankai University, China.

Dongkai Yan 閻東凱 is Lecturer at the College of History and Civilization, Shaanxi Normal University, China.

Li Zhang 張莉 is Associate Professor at the Northwest Institute of Historical Environment and Socio-Economic Development, Shaanxi Normal University, China.

Marlon Zhu 朱瑪瓏 is Assistant Research Fellow at the Institute of Modern History, Academia Sinica, Taiwan.

Introduction

Ts'ui-jung Liu

Environmental history has gradually evolved into a vibrant and well-established historical subfield since its emergence in the 1970s.[1] It has broadened the horizons of historical research, beyond mere human affairs, to include the study of human beings and their interactions with natural and man-made environments. At the same time, this broadened scope has attracted more scholars from different disciplines to join the echelons of environmental history researchers. In recent years, this has been notable in relation to the environmental history of East Asia.

Since the establishment of the Association for East Asian Environmental History (AEAEH) in late 2009, two biennial conferences have been held and an edited volume of papers, revised after the first conference, has been published.[2] This new volume includes some of the papers presented at the second conference held in October 2013.

Thematically, this volume focuses on the local realities and environmental changes in East Asia. Every locality in the world, regardless of its geographical size and political status, has a unique cultural and historical heritage, social, economic, and political realities, as well as natural and resource endowments. With the availability of ample historical documents on East Asia, scholars can select suitable materials with which to narrate the environmental histories of their chosen localities. In this way, scholars engaged in environmental history research in East Asia, as in other regions of the world, have been adopting an interdisciplinary approach.

This volume consists of three parts. The first part includes two chapters that deal with the thoughts and attitudes toward nature in China and Japan, respectively. The second part consists of four case studies relating to three regions: the Lower Yangtze Delta, Northwest China, and Taiwan. The final part of the volume contains two chapters on the Tokyo area: the first one on a volcanic eruption and the second one on the construction of modern bridges and roads. The time frame of these eight chapters spans the fourth century BC up to the present.

In Chapter 1, Lihua Wang presents a soundscape study with a focus on "listening" to the sounds of bamboo. Wang utilizes records from Chinese literature, painting, music, philosophy, architecture, and history to reveal the sounds conceived by the Chinese literati from the ancient period to the pre-modern period. Against a background of the different sounds of nature and human society, this chapter presents a subtle narrative on the sounds that surround bamboo groves. It leaves the reader with a vivid sense of the lingering sounds of bamboo. This is perhaps the first study that presents a soundscape within East Asian environmental history.

In Chapter 2, Satoshi Murayama elaborates on the concept of an "industrious revolution," coined by Hayami Akira, to describe the labor-intensive technologies developed in Tokugawa Japan. Murayama focuses his discussion on livestock production and its environmental historical implications during Japan's Tokugawa period (1603–1868). He demonstrates that, along with an increase of the population during this period, there was an increase in livestock production in many Japanese villages. He further compares the case of Japan with that of early modern commercialization in Europe studied by Jan de Vries. Moreover, he raises some questions for further study on the nature of the relationship between an "industrial revolution" and "industrious revolution," and the implications of an "industrious revolution" for interpreting human–nature interactions.

The next section of the volume consists of four case studies concerned with the utilization of land and water resources in mainland China and Taiwan.

In Chapter 3, Cho-ying Li analyzes Xia Yuanji's water management in the Lower Yangtze Delta from 1403 to 1405. In this particular case, Li aims to investigate the relationship between the hydraulic control and the legitimization of Yongle's reign (1403–1424) during the Ming Dynasty (1368–1643). Li uses ample primary documents and secondary literature in constructing his narrative and analysis. His chosen methodology is the most appropriate one for this case study. It involves an exploration of the political expectations, cultural implications, limitations, and possibilities inherent in existing environmental conditions, and the combined effects of all of these factors. This case study aptly illustrates the relationship between hydraulic control and political ecology in Chinese history.

Chapter 4 also focuses on the Lower Yangtze Delta, but the time frame of this study is the late nineteenth century. Marlon Zhu examines the records of the *North China Herald*, a commercial press in Shanghai, to explore the "Woosung Bar" controversy. This issue pertains to the dredging of the Wusong River since the 1870s. Various actors within Shanghai's commercial community, including shipping companies, banks, commercial firms, the Maritime Customs, as well as Chinese and foreign diplomatic authorities, were involved in the controversy. Zhu argues that the media in China during the late nineteenth century played a considerably influential role in relation to the environment and created a discursive space for the articulation of conflicts among competing commercials interests, and between the Chinese and British states at that time.

Chapter 5 focuses on the Dahaizi Reservoir in the Manas River Basin in Xinjiang, Northwest China. In this chapter, Li Zhang and Dongkai Yan examine the dispute that occurred over the expansion of this reservoir during the 1950s. They analyze four issues by utilizing detailed written records (some unpublished) that are available on the debate surrounding this project. These issues are (1) the land reclamation and frontier consolidation in Xinjiang, (2) China's changing domestic situation, (3) changes in the Sino-Soviet relationship, and (4) changing attitudes toward the impacts of water projects. They point out that, even though policymakers were aware of the environmental damages associated with the reservoir's expansion, they nevertheless decided to sacrifice the environment for the sake of economic prosperity. Evidently, this genre of decision-making is not unique to this case, but pertains to many cases of modern economic development across the globe.

In Chapter 6, Chun-Chieh Chi focuses on the development of aboriginal communities in Hualien, Eastern Taiwan. Chi analyzes how indigenous people in this region have tried to make their communities more ecologically friendly, revive their traditional culture, and attract tourists. He also examines how non-governmental organizations have interacted with indigenous communities to make their development projects more sustainable. Moreover, he points out some of the ways in which tourism development has helped aboriginal communities to preserve their traditional ecological knowledge.

The last two chapters are concerned with human responses to the natural disasters that have occurred around the Tokyo area since the eighteenth century.

In Chapter 7, Beatrice M. Bodart-Bailey examines the events that occurred around Mount Fuji's volcanic eruption in 1707. She uses numerous primary records to reconstruct the actual situation of the eruption, the relief works carried out after the disaster, and the incomplete reconstruction works initiated by Tokugawa Tsunayoshi (1646–1709). She points out that, after Tsunayoshi's death in 1709, the failure of the *bakufu* to implement his policies incurred hardships for the common people and resulted in social upheaval. The widespread disturbances that followed hastened the fall of the Tokugawa regime.

Chapter 8 focuses on the construction of modern bridges and roads across the Tama River since the Meiji period (1868–1912). M. William Steele traces, in detail, the history of the construction of modern bridges and roads for crossing the Tama River against a background of frequent occurrences of flood and earthquake around the Tokyo area. These projects were initiated by the Japanese government, first with the help of Western specialists, and subsequently by Japanese civil engineers trained in Western technology. Steele asserts that ferro-concrete construction, which prevailed from 1918 to 1932, left a powerful, albeit ambiguous, legacy on postwar Japan.

In sum, this new volume highlights the recent studies on East Asian environmental history by scholars of different disciplines who have applied rich historical materials, maps, images, and statistical data to reveal the local environmental realities infused by global perspectives. As there are still relatively few publications on East Asian environmental history published in English, this volume can be used as a source of reading materials for related university courses.

Acknowledgments

I would like to thank Academia Sinica and the National Don Hwa University for providing support to the Second Conference of East Asian Environmental History held in October 2013. Academia Sinica provided funds for the Conference, and the National Dong Hwa University provided the venue. Many scholars have helped in reviewing the papers; their contributions are gratefully acknowledged.

Notes

1 For a recent overview of this field, see J. R. McNeill, "The State of the Field of Environmental History," *Annual Review of Environment and Resources*, 35 (2010): 345–374; available online: www.annualreviews.org/doi/full/10.1146/annurev-environ-040609–105431#, accessed on October 1, 2014. This article covers the works of historians, archeologists, geographers, and others in various regions across the globe, but places heavy emphasis on South Asia and Latin America. See also Bao Maohong 包茂紅, *The Origins of Environmental History and its Development* (環境史學的起源和發展), Beijing: Beijing University Press (2012): 3–236. This book traces the development of environmental history in America, Latin America, Britain, Africa, India, Southeast Asia, Australia, China, and Japan in nine chapters, in addition to one chapter on environmental historiography and the other on new trends in international environmental history.

2 Ts'ui-jung Liu (ed.), *Environmental History in East Asia: Interdisciplinary Perspectives*, London and New York: Routledge (2014): xiv+400 pages. It should be noted here that the format of style is followed by this new volume in that Chinese and Japanese names mentioned in the texts and notes have been romanized with the surname preceding the first name, without the use of a separating comma. However, the names of authors of each chapter have been arranged according to the accepted Western format. Also, for avoiding redundancy, the romanization of the article and book titles cited in each chapter is only provided in the bibliography instead of in both endnotes and bibliographies.

Bibliography

Bao Maohong 包茂紅. *Huanjing shixue de qiyuan he fazhan* 環境史學的起源和發展 [*The Origins of Environmental History and its Development*], Beijing: Beijing University Press, 2012, pp. 3–236.

Ts'ui-jung Liu (ed.). *Environmental History in East Asia: Interdisciplinary Perspectives*, London and New York: Routledge, 2014, xiv+400 pages.

McNeill, J. R. "The State of the Field of Environmental History," *Annual Review of Environment and Resources*, 35 (2010): 345–374; available online: www.annualreviews.org/doi/full/10.1146/annurev-environ-040609–105431#, accessed on October 1, 2014.

Part I

Thought and attitude toward nature

1 Listening to bamboo

The Chinese literati's attitude and behavior toward the sounds of nature in the pre-modern era

Lihua Wang

Introduction

Our planet is full of all kinds of sounds. Nature and man produce different sounds in every second. The sounds of different wavelength and frequency impact on our physiology and psychology in various ways and to different extents. Some are pleasant, some disturbing, and thus, some we like, some we detest. All sounds, no matter natural or man-made, as long as they have an impact on people, constitute a part of the human environment. Today, noise pollution (for which mostly humans are to be blamed) has become one of the environmental problems we are facing. At the same time, many beautiful sounds of nature have been fading away and we have been losing a lot of auditory happiness. For these two reasons, environmental historians need to pay more attention to the sound history and take it seriously as a subject in the exploration of the historical relationship between man and nature. We need to know how noise pollution gradually came into being, so as to find a solution to the problem. Moreover, in order to locate this problem within a broad historical context, we need to understand how humans have perceived and treated different kinds of sound. What knowledge, ideas and aesthetic consciousness did humans generate and develop?

Historically speaking, human cognition, attitude and feelings toward sound, and even sound itself have been changing. Consciously or not, some memories of sounds have been preserved in historical records. This provides some possibilities for the study of related problems from the perspective of environmental history. To extract these historical memories is a very worthwhile exercise that will also help to resuscitate a sense of natural beauty and to create a healthy acoustic ecology and a pleasant soundscape.

But unlike a concrete matter, sound never stays. Compared with visual memory, it is far more difficult to retain echoic memory and to bring it alive. This might be the reason why traditional historians so far have never paid any attention to problems such as how humans perceive sound, how sounds of the nature as an environmental factor that is non-figurative and non-static have gradually entered into human culture. Research on this topic is almost non-existent in environmental history studies too. Individual western environmental historians seem just to have begun to explore this field.[1]

However, research in other fields may give us some inspiration. Some scholars of literature studies occasionally touched on the problems of acoustic images, descriptions and their associated meanings in classical Chinese poetry and prose. Recent studies on gardens and architecture discuss soundscapes created by ancient Chinese.[2] But their approaches are different from that of environmental historians. Behaving as "the first man to eat crabs," I take "listening to bamboo" as a case study for a first, rough investigation of the traditional Chinese literati's attitude and behavior towards the sounds of nature.

Sounds of nature enter into culture: the ways of reason and sensibility

The ancient Chinese did not have acoustics in the modern sense (music might be an exception). But they were masters in cognizing and describing the sounds of nature based on experience and emotion. They have left in their poetry, painting and architecture countless wonderful images and descriptions of soundscape. They even have initiatively created many pleasant acoustic environments.

The ancient Chinese language has a rich vocabulary to express, describe and classify different sounds that reflects to some extent people's knowledge and attitudes. For instance, Zhuangzi (莊子, *c.*369–286 BC) classified sounds into the three types of "sounds of human" (*renlai* 人籟), "sounds of the earth or the land" (*dilai* 地籟) and "sounds of the heaven" (*tianlai* 天籟). Among the sounds, "*tianlai*" is the most beautiful and this term can also be used to describe the most beautiful music and songs. Whereas the disturbing noise in human societies and cities was called "earthly bustle" (*chenxiao* 塵囂). But there are also some annoying, fearful and sad sounds in nature, of course. The chirring and rasping of cicadas in the hot summer days, the wild goose honks in the autumn sky, the crow caws on the cottage roof in winter, as well as the wolf howling in the wild, the ape crying in the three gorges of the Yangtze River, etc. In the literary works, most of these have certain aesthetic implications.

Basically, the sounds described in the ancient works may be divided into two groups: the nature sounds and the sounds of human society. The former includes sounds of wind and rain, horses and birds, and insects such as orioles, cuckoos, partridges, pelicans, bees and cicadas. The latter includes the sound of the Qiang flute (*Qiangdi* 羌笛), the vertical flute, the trumpet, bell, drum, bugle, flail, and the sound of beating and washing clothes (*daoyisheng* 擣衣聲) in the deep night. The description of sounds in ancient Chinese was mostly placed on or set off certain emotions and thoughts.

Sometimes the writers attributed certain feelings to a sound, such as "the apes cry sadly" (*yuantibei* 猿啼悲), "the birds sing with resentment" (*niaotihen* 鳥啼恨) to reflect the poet's subjective feelings. Sometimes, sound let them long for the past, think of their lovers or feel the vagaries of life. Some descriptions were just artistic devices to render the atmosphere, though without any specific meaning, such as in "listening to the rain and wind all night in a small building" and "it rains foggy while the cuckoo laments."[3]

To express deep-hearted feelings by describing sounds of nature was in fact a very old literary tradition with a history of at least three thousand years in China. Both big sounds from the thunderclap and the roaring of storms to phoenix tweets, tiger growls and dragon whistles, and small sounds, such as the nightingale jugs or the swallow twittering in spring, the cold cricket singing in the late cold winter, or small water droplets dropping into a secluded spring valley, the cold breath of autumn shaking down the leaves, all kinds of natural sounds gradually entered into our culture, and the history became animated.

Rich descriptions of sounds can be found in the first anthology of ancient Chinese poetry, *The Book of Poetry* [*Shijing* 詩經], which is believed to be compiled by Confucius (551–479 BC). The earliest descriptions of the sounds of thunder, wind, rain, as well as the sounds of many kinds of birds, insect and mammals can be found in these poems. In "The Song of July" (*qiyue* 七月) for instance, sounds of nature changed in accordance with the seasonal changes. From July to October, crickets sing in the wild field, under the eaves, in the door and under the bed separately. All sounds co-constructed the lively and season-changing acoustic environment for the ancient Chinese farmers in the Bin area (豳, in the central Shaanxi province today).[4] The descriptions in *The Book of Poetry* did not only record the ancient Chinese's original ecological knowledge about sounds of nature, but also reflect their complex emotions through their awareness of the nature such as joy, longing, resentment and sadness. *The Book of Poetry* can be regarded as the first precedent for the interaction between the Chinese inner-heart voice and sounds of nature; and it had a profound cultural influence.

In the later history, there were a lot of masters who were good at description of the sounds of nature and there were abundant wonderful works. Qu Yuan (屈原, *c.*340–278 BC) for instance, by only four sentences and 28 characters in his poem of the Mountain Fairy (*shangui* 山鬼), vividly demonstrated a soundscape including the thunder and lightning storms, ape wail, tree leaves rustle, and expressed people's bitterness and nostalgia induced by the soundscape.[5] Since the Han Dynasty (206 BC–AD 220), there is a constant growth in techniques and methods of describing sounds of nature. For instance, *Hanfu* (漢賦, a type of descriptive prose interspersed with verse) included a large number of audio descriptions. In the middle ancient times, sound descriptions by the poets had reached a very subtle degree. For instance, Wang Ji (王籍, a poet of the Liang Dynasty, 502–557) once wrote, "the forest is more static in the chirping of cicadas, the mountain is more secluded with bird songs."[6] Wang Wei (王維, (699–759 or 701–761), a great poet and painter of the Tang Dynasty (618–907), wrote in a poem entitled "bird sings in the deep valley" (*niaomingjian* 鳥鳴澗) that "birds are frightened as the moon climbs the mountain, and they tweet in the deep stream valley now and then."[7] These descriptions not only show the natural realms of spaciousness through sound description, but also show that the "dynamic" and "static" in the natural world are not absolutely opposite, they are also complementary in certain circumstances, i.e., sound occurs in static and appears more static and quiet!

History is an academic discipline that is mostly rational rather than perceptual. The "scientific meanings" of sounds in the natural world that the ancient Chinese described are the more important aspect to be explored from the perspective of an environmental historian. What I would like to point out first is that: listening to sounds of nature once was a very important or even a major way to grasp the environmental changes (here referring to the seasonal variation of nature). Abundant descriptions in ancient poetry and prose induce us to meditate on the seasonally changing sounds of nature, on which the ancient Chinese determined their work and living schedules in the four seasons in a year. The Chinese have already noticed the seasonal changes of nature sounds since very ancient times. They recorded them as what we today call "phenology" (*wuhou* 物候, natural phenomena that recur periodically), marks of time moving, and gradually developed a phenological calendar—Xiali (夏曆) which is the earliest Chinese lunar calendar and believed to have been created in the Xia Dynasty (*c.*2206–1766 BC). The "*Xia Xiaozheng* (夏小正)," a chapter in *Dadai's Book of Rites* [*Dadai liji* 大戴禮記], was its textual basis. Among the numerous phenological phenomena that mark seasonal changes, the sounds of chirping and crying animals were of special importance. Since very ancient times, in both official and private writings, and even in folk proverbs, animal calls and other sounds of nature are important references to determine seasonal change. Some knowledge and customs have been passed down and are still alive today. This is a very interesting phenomenon in Chinese environmental history worthy of a systematic exploration.

The real history, however, often abounds with sensibility. The ancient Chinese people used to describe the sounds of nature mainly relying on their perceptual rather than rational cognition. Since the natural environment is constantly and periodically altering with seasonal changes, human emotions also tend to alter seasonally. The spring longing, summer fret, autumn sorrow and winter groan of the ancient people were different complexes due to seasonal environment changes. All of those emotional changes were due not only to the seasonal changes of the color, but also to the changes of the sounds of the environment. Furthermore, they were often expressed by means of sound descriptions.

For example, ancient Chinese scholars frequently expressed strong autumnal melancholy.[8] Every year when autumn is coming, the air gradually became cold, nature images and sounds also changed. The atmosphere was dominated by chilly drizzle and blowing wind, withering yellow leaves and grass, the hoarse chirping of the cicada and the sorrowful cry of the wild goose, all of which induced a sense of vastness, desolation and bleakness in humans. For thousands of years, countless men of letters have continued to chant and paint such soundscapes. A typical example is Ouyang Xiu (歐陽修, 1007–1072), a leading scholar in the Northern Song Dynasty (960–1126), who wrote "*The Prose poem of Autumn Sound*" (*Qiushengfu* 秋聲賦) to express autumnal melancholy through sounds. It had a deep influence on later literature, calligraphy and painting. When reading the prose, the reader would feel he himself is personally in the place described by Ouyang, and spontaneously generate complex feelings of bleakness, anxiety and grief, etc.[9] Since then the literati either copied the article by their own calligraphy, or drew paintings inspired by its artistic conception. These can be observed by the works of Zhao Mengfu (趙孟頫, 1254–1322), Dong Qichang (董其昌, 1555–1636), Ren Bonian (任伯年, 1840–1895), Yuan Songnian (袁松年, 1895–1966) and Li Kuchan (李苦禪, 1899–1983). It is notable that although the painting of Li Kuchan is titled as "Listening to the Rain" rather than "Listening to Autumn Sound," the residual banana, the thin bamboo and the cold heron clearly express the chill of autumn.

Sounds of nature exist everywhere, any time. Knowledge, ideas, emotions and attitudes that are produced by the perception of sounds of nature change from time to time. In fact, both the perceiver and perceived have been changing constantly with the development of history and this reflects one aspect of the complexity of the relationship between man and nature. This is impossible to be discussed completely in a single paper. Here the author only offers a preliminary investigation of the case of bamboo.

Descriptions of sounds in bamboo groves: text and image

An extensive literature search reveals the very interesting fact that Chinese people began very early to record sounds of nature, including those related to animals and plants. In early periods, however, descriptions of plant-related sounds were notably fewer than those of animals and literary records of sounds in bamboo forests appeared even much later. The reason was obviously not that the early people did not pay attention to the sounds in bamboo groves, but that the geographical center of early literary records was the north of China where bamboo forests were far less abundant than in the southern part of the country.

During the period of Six Dynasties (*c.*220–588), along with the geographical discovery in South China, the sounds of bamboo gradually entered into the historical narrative. The *Gazetteer of Jingzhou* [*Jingzhou ji* 荊州記] written by Sheng Hongzhi (盛弘之), who lived from the late fourth to the early fifth century in Jingzhou (in south Hubei province today), recorded that: There were a giant breed of bamboo and some other smaller ones in the Mount Linghedong (臨賀冬山). When the wind blew, the bamboo produced sounds just as *xiao* (簫) and *guan* (管) (two ancient types of flute). Approximately in the same period, the *Gazetteer of Yongjia* [*Yongjia ji* 永嘉記] also recorded that at Yongjia (in southeast Zhejiang province today), there was a stone plateau of ten square meters which was named Fairy Altar (Xiantai 仙台) on the summit of Mount Yangyu (陽嶼山). Four bamboos of the Jinzhu (筋竹) family grow in luxuriant green nearby. When the wind blows, the bamboo spontaneously generates music.[10] These records were full of magical color; whether true or not, they reflected at least that the scholars had noticed the sound of bamboo.

There are only very few poems on the theme of bamboo in the period of Six Dynasties. Poets occasionally mentioned bamboo sounds but without any detailed description. This indicates that they were still lacking a subjective consciousness to appreciate bamboo sounds. We have not found any notable change of this in the poetry of the early Tang period. However, the situation was very different in the later Tang poetry and prose; descriptions of bamboo sound increased significantly. The poets creatively used various metaphors to describe different kinds of sounds in bamboo groves.

Linhu Chu (令狐楚, 766 or 768–837), for instance, was the first to compare the sound of wind blowing through a bamboo grove to the sound of rain.[11] Liu Yuxi (劉禹錫, 772–842) took the lead in describing the sound of wind blowing through bamboo groves as being like the sound of Yaochi (瑤池), the most beautiful fairyland where the Queen of the Western Heaven (Xiwangmu 西王母) lived according to ancient Chinese legends.[12] Yuan Zhen (元稹, 779–831) wrote several poems with different descriptions of sounds in bamboo groves and compared them either to the most beautiful music or to a frightening sound that even the ghosts feared. He also mentioned birds singing and insects buzzing in bamboo groves.[13] Yuan Zhen's best friend, Bai Juyi (白居易, 772–846) was a lifetime bamboo fan who tops the list of the bamboo description masters in the Tang Dynasty. Bai's evocations of the sounds of bamboo were the most ingenious. For example, he described, on hot summer days, the whistling sounds of the bamboo grove surrounding the house, nearby the window, that brought about coolness just like the gently blowing autumn wind.[14] He was also the first to describe the sound of heavy snow breaking bamboo.[15] Poets in later periods adopted this as a typical image of quiet and severe cold in winter. Another poet, Xue Neng (薛能, *c.*817–880), described a different quiet image of forest-bamboo shoot sheaths.[16]

After Bai Juyi, the poetic observation of bamboo sounds became increasingly careful, in the greater detail, concreteness and vividness of the descriptions. The sounds of bamboo groves have been compared to flowing spring waters by Jia Dao (賈島, 779–843),[17] dragon chants by Duan Wengui (段文圭, a poet of the late Tang Dynasty)[18] or the sounds of drying amur silver-grass by Tang Yanqian (唐彥謙, ?–93).[19] Moreover, Xu Yin (徐夤, a poet who lived in the late Tang Dynasty) once declared that there were six different tonalities in the bamboo grove.[20] Under the writing brush of Liu Jian (劉兼, living around 960), the sounds of wind knocked bamboo like "breaking jade" (*suiyu* 碎玉) and that increases quietness instead of noise.[21] Every spring, bamboo rhizome scurry and countless shoots grow competitively when the first thunder sounds in the drizzly rain. The whole bamboo forest is full of energy of new creation.[22]

It is worth noting the influence of Bai Juyi's creations of descriptive metaphor. Many poets of later generations accepted that autumn breaths existed in bamboo groves. It gradually became very common to link sounds in the bamboo grove to the spirit of autumn, wind, rain and coolness. When a breeze gently shakes the bamboo, it brings about wisps of coolness and it seems like autumn is coming early. Gentle sounds occur as wind waves bamboo branches and leaves, just like the sound of autumn rain which can drive the summer heat away. Many related descriptions can be found in the Tang poetry. Li Qunyu (李群玉, 813–860), Wu Rong (吳融, 850–903), Liu Yanshi (劉言史, 742–813), Qin

Taoyu (秦韜玉, ?–), Tang Yanqian, Li Zhong (李中, living around 942) and many others described the bamboo grove in one way or another. All the descriptions demonstrated a common imagery that "bamboo grove sounds convey the meanings of autumn." But please be aware: it was a mood completely different from the sad feeling of the decay in autumn we mentioned above. In the opinion of Tang poets, bamboo groves sounded pleasantly because they could bring about autumn coolness to the annoyingly hot summer days.

It was because of this positive evaluation that the Tang poets enjoyed listening to bamboo. Li Zhong once said in a poem that he would go to a bamboo grove in autumn when he wanted to listen to the rain. Zuo Yan (左偃, living around 940) recalled in a poem that when he was living with a monk ten years ago, they stayed up all night to listen to the rain in a bamboo room. It was in the late period of the Tang Dynasty when "listening to bamboo" began to become an initiative behavior of appreciating nature sounds. It gradually developed into a refined pastime of the literati group and its influence lasted for more than a thousand years.

From the Song Dynasty, the literati observed bamboo sounds in yet more detail. Their methods of description became increasingly sophisticated. Descriptions of all sounds, such as the sound of wind, rain and the dropping of shoot sheaths in bamboo groves can be easily found in many collections of poems and prose. There were more and more creative expressions. For instance, Su Shi (蘇軾, 1037–1101), a leading literati in the Northern Song Dynasty, once compared the sound of the wind blowing through bamboo groves to that of swords colliding with helmets and armor.[23] In the Yuan Dynasty, Zhou Di (周砥, living around 1367) and Ma Zhi (馬治, living around 1367) enjoyed and described the bamboo nearby a pond and compared the sounds produced by the wind blowing through bamboo groves late in the night to "phoenix songs" (*fengming* 鳳鳴), "dragon chants" (*longyin* 龍吟), the "sounds of the heaven," the "sounds of metal stone instruments" (*jinshizhisheng* 金石之聲) and the "sounds of waves" (*taosheng* 濤聲), etc. They enjoyed the quietness in the sounds of wind blowing through the bamboo and entered into "the realm of Zen" (*chanjing* 禪境), a miraculous condition of forgetting words as the meaning has been caught (*deyi wangyan* 得意忘言).[24]

In the Ming Dynasty (1368–1643), Shen Zhou (沈周, 1427–1509) once compared the sounds of bamboo to "the gentleman's words." But it was more common to compare it to the lingering charm of wind-blowing fairy jades (*fengchui biyu* 風吹碧玉) in the period. Clearly the sound of bamboo was never perceived as noisy. It made people feel easy, calm and happy.[25] In fact, since the time of Confucius, bamboo has been often likened to jade. No wonder Shen had another poem about listening to bamboo but entitled "Listening to Jade" in which he used "*green langgan*" (琅玕, a pearl-like stone), "*congzheng*" (琮琤, jangling of jade), "*linlang*" (琳琅, beautiful jade) and "*huanpei*" (環佩, jade pendants worn on a girdle) to describe the sounds of wind blowing through bamboo groves.

Another poet, Shen Shixing (申時行, 1535–1614), wrote in a poem entitled "Breaking Bamboo in Heavy Snow" also describing "wind blowing through bamboo" (*feng chuizhu* 風吹竹) as "wind knocking at jade" (*feng qiaoyu* 風敲玉). Furthermore, in their drawings and painting poems, some of the Ming literati also entitled their bamboo works "Listening to Jade" (*tingyu* 聽玉). For instance, Cheng Wen (程文, ?–?) once painted a picture entitled "Listening to Jade." According to a poem on a painting of his friend Wen Zhengming (文徵明, 1470–1559), it actually reflected a leisurely and quiet life represented by enjoying the sounds in the bamboo grove.

The poem of Wen Zhengming depicted a circumstance full of Zen meanings (*chanyi* 禪意): In a quiet cottage under the shadow of bamboo groves, the lord lived with bamboos, listened to the bamboo sounds and enjoyed a state of deep silence. He did not listen to the bamboo with his ears, but with his heart. He communicated with the bamboo in silence, felt what the bamboo felt and "forgot both bamboo and man himself" (*renzhu liangwang* 人竹兩忘). Cheng's painting is not extant now. But the realm described in Wen's poem can be roughly imagined through the two pictures by Xia Chang (夏昶, 1388–1470) and Tang Yin (唐寅, 1470–1524) respectively; both were famous painters in the Ming Dynasty.

It is interesting to mention that the Ming literati had some very unique imagination of the sounds in bamboo groves. For example, Li Rihua (李日華, 1565–1635) mentioned in his diary that he once came across in the market a bamboo painting of Tang Bohu (唐伯虎, i.e., Tang Yin). The poem on the painting said:

> A grove of bamboo surrounds the mountain house shadowy, in the autumn night the sounds of the bamboo can be heard as thickly as rain; sometimes the bamboo sounds like the noise of silkworms sucking mulberry leaves and sometimes it sounds like crabs secretly crawling on sand.

Nobody could have this kind of imagination except gifted scholars in Jiangnan (江南) like Tang Yin.

Compared with poetry, prose has fewer metrical constraints and can describe the bamboo sound more fully and concretely. There were plenty of related works in the history. For instance, Zhao Mengfu, a famous painter of the Yuan Dynasty, depicted the fun of listening to bamboo in his "Xiuzhu Fu (修竹賦)," a prose poem of beautiful bamboo. He described a complex soundscape with sounds of wind knocking, rain showering, heavy snow pressuring, as well as cranes whooping and cicadas calling. All the sounds co-composed a natural symphonic music just as if there were various instruments performing together. The listener felt like returning to the very beginning of the creation of the world, enjoyable and pleasant.[26] As Zhao Mengfu had clearly mentioned, there were not only sounds of wind and rain, but also sounds of animals in the bamboo grove. The bamboo groves belonged not only to humans. Birds, insects and other animals were also residents or visitors. They produced sounds that could be heard in the groves. Man and animals formed a community of life together.

It is almost impossible to offer a comprehensive list of the images that the Chinese ancients used to describe the sounds of bamboo. What we can say is only that the bamboo sound perception through intuitive and empirical observation of ancient literati has reached an incredible detail. They used a variety of literary rhetorical methods such as the symbolic and anthropomorphic, and contrasting to describe all kinds of sounds in bamboo grove. As a result, the ancient Chinese literati have fully demonstrated their understandings of natural rhythm, creatively described many quiet, elegant and pleasant soundscapes, expressed their special natural aesthetical preferences and taste, and all these reflected their consciousness of the life world.

Yet, the most intuitive way might be painting. Since the Song Dynasty, as the growing and maturity of painting, chants of bamboo sounds from the literati have not been limited to poetry and prose; they also intended to represent bamboo sounds through painting, generally with painting poetry as an important supplement, of course. In their views, it was the highest artistic level if a painter could make the appreciator as if hearing sounds from his painting. In Chinese traditional painting, what the artists wanted to express was in fact not just their perceptions of landscapes and colors, but also what they heard, felt, thought of the nature sounds.

As we have mentioned above, Zhao Mengfu was very interested in bamboo sounds. That was reflected not only in his prose, but also in his paintings. As a very famous bamboo-painting master, he intentionally or unintentionally expressed his sound images of bamboo groves with quite a few of his works. When looking at his "Secluded Bamboo and Hoopoe (幽篁戴勝圖)" and "A Self-portrait (自畫小像)", it seems as if they can let us hear the birds pleasantly singing and the bamboo whispering when gently shaking branches and leaves. In fact, we can find plenty of paintings with this kind of artistic intention from the gallery of the Song Dynasty to Modern times.

Listening deep to one's heart: the connotation of "listening to bamboo"

"Listening to bamboo," including listening to the sounds produced by the wind, rain and animals, etc. in the bamboo grove, was initially an unintentional behavior of only a few persons. Roughly speaking, "listening to bamboo" as an elegant fashion in the literati society was not been formed until the Tang Dynasty. But it already had become a popular hobby not later than the Yuan Dynasty. As more and more scholars took "listening to bamboo" for joy, it gradually evolved into a consciously pursued elegant lifestyle or fashion, a common behavior that may not be easily understood today.

The Song Dynasty was a very interesting epoch in which the custom of planting and admiring the beauty of flowers prevailed. Even men liked wearing flowers on their head for beauty. Planting and chanting bamboo was also popular, the custom of initiatively "listening to bamboo" also started in this period according to the relevant poetry. Wang Anshi (王安石, 1021–1086)

seems to be the first person who put "listening to bamboo" on a par with "admiring flowers." In a poem he said: "I feel calm when admiring flowers and listening to bamboos, the wind blowing in the bamboo grove wakes me up from drunkenness."[27] But using "listening to bamboo" as a title of a poem was initiated by Lu You (陸遊, 1125–1210), one of the most famous poets in the Southern Song Dynasty. In Lu's opinion, enjoying the sound of rain in a deep bamboo grove could help one to sober up, feel healthy and relaxed, and even recover from illness without medicine.[28] He obviously took "listening to bamboo" as a great joy. Although the joy of "listening to bamboo" was occasionally mentioned by Wang, Lu and others, it does not seem to have been a common behavior among the literati until the Yuan Dynasty.

The situation was obviously different in late imperial times when poems and essays entitled "Listening to Bamboo" became more and more common. According to their descriptions, people often enjoyed listening to the sound of wind or rain in the bamboo grove from different places such as verandas, pavilions, study dens, sunken flower-beds, or nearby windows, streams and springs. Many elegant figures entitled their buildings or resting places "Listening to Bamboo." All this indicates that listening to bamboo had become a kind of lifestyle fashion that people pursued consciously.

In addition to listening to the sounds of bamboo wind and rain, the ancients also listened to "firmiana rain" (*wuyu* 梧雨), "banana rain" (*jiaoyu* 蕉雨), "lotus rain" (*heyu* 荷雨) and to "pine wind" (*songfeng* 松風). Rain drops on firmiana leaves (*yuda wutong* 雨打梧桐), rain beats banana leaves (*yuda bajiao* 雨打芭蕉), raindrops hit the broken lotus (*yuda canhe* 雨打殘荷) in autumn and wind roars in pine forest (*fenghou son-lin* 風吼松林) were all important natural aesthetic images in pre-modern China.

Listening to "firmiana rain," "banana rain," "lotus rain" and "pine wind," however, were all not as quiet and appropriate as listening to bamboo. Wu Yulun (吳玉綸, 1732–1802), a scholar in the Qing Dynasty once commented:

> Among all the plants, bamboo possesses knots [*jie* 節, a symbol of moral courage] and accommodates to frost and snow, its shadow fits the moonlight, its figure fits mist, its sound fits wind and dew, particularly fits rain. When we tell the beauty of all things, shape is not as important as sound; when observe things and get insights, watching is not as good as listening.... The sparse rain is suitable for firmiana tree, the rapid and intensive rain is suitable for pine tree, and both of them are suitable for bamboo. Thus bamboo grove is especially suitable for us to listen to the rain. It is a matter related with disposition. The one who does not listen cannot understand, the one who listens also cannot understand.[29]

It was for this reason that most scholars of the Ming and Qing Dynasties considered that man could not listen to wind and rain without bamboo. The main purpose of planting bamboo nearby houses was to obtain the joys of "listening to bamboo." We can find plenty of poems expressing this kind of intention. For instance, Song Ne (宋訥, 1311–1390), an early Ming poet, said that: “Planting pines is to invite the clouds to stay, planting bamboo is for listening to the sound of rain.”[30] Qian Qianyi (錢謙益, 1582–1664), a famous scholar at the turn of the Ming and the Qing Dynasties, said: “The sound of wind is mostly to be listened to in the bamboo grove, and the moon cannot be appreciated without pine trees.”[31] Shen Shourong (沈壽榕, 1823–1884) said: “Transplant a few poles of bamboo to share the joy of listening to wind and rain.”[32] And Fa Shishan (法式善, 1752–1813) said: “You need to plant some bamboo if you want to listen to the wind and rain, you would dislike watching fishes if you are not the like of Huizi and Zhuangzi.”[33] It seems that an exception could be Zhang Yingchang (張應昌, 1790–1874), he was awarded a teaching job in Huichang 會昌 county, Jiangxi province in 1558), as he once questioned the necessity of a thousand poles of bamboo for listening to wind.[34] Such a single different tune, however, could not deny the bamboo's importance for listening to wind and rain; on the contrary, it just reconfirmed a common historical truth that the elegant literati loved to “listen to bamboo.”

The more noteworthy aspects were not the Ming and Qing literati's clever, vivid and wonderful accounts of the scenes and feelings of listening to bamboo, but their interesting ideas and implications about nature, life, music, etc. through philosophical explorations and Zen allegories/Buddhist subtleties (*chanji* 禪機). Some discussions might look like word games, with some slight tastes of “wild-fox Zen” (*yehuchan* 野狐禪), or non-orthodox teachings, but nobody could totally deny their significance and value whether in view of acoustic ecological aesthetics or the philosophy of life.

The ancients described and understood the sound of bamboo to a surprising level of ineffable depth. They listened to bamboo, heard it with their ears, felt it with their hearts, it entered their spirit, and thus they attained enlightenment, forgetting their body, and at last reached a mysterious realm of “harmony between heart and environment” that can only be sensed and is impossible to be explained in words. Our basic impressions are that “listening to bamboo” embodied a cultural spirit of respect for nature, being close to nature, that reflected the literati's inclination to an elegant, quiet and leisurely life; and that for the literati, “listening to bamboo” was not only an entertainment, but also a spiritual pursuit. It was actually listening to one's inner heart. The key lay in the listener's mood or state of mind. The “sound environment” of the bamboo grove only provided an opportunity or a medium to the self-spiritual insight of the literati, just like Wen Zhengming had said: “Who says the sound exists in the bamboo? Whether you could hear it or not merely depends on yourself.”[35]

The “opportunity” or the “medium” was very important, of course. In the noisy world, the sounds of the bamboo grove could help the listener to calm his upset mood, sink into deep thinking, rise to exquisite feelings, temporarily put aside the desires for pursuit and competition in the “Vanity Fair,” and throw away the disturbing thoughts. In a quiet circumstance, the listeners could devote themselves to the sound of the bamboo grove, communicate with nature, so as to obtain a moment of spiritual liberation and return to the freedom and innocence of themselves. Because of listening to bamboo, their worries were eliminated, their minds renewed and diseases were alleviated or even cured.[36]

The ancient Chinese created several acoustic images such as "listening to the rain in the bamboo grove," "listening to the wind in the bamboo grove," "the autumn sounds in a bamboo grove" and so on, containing some particular eco-aesthetic meanings. They are still worthy of appreciation and taste today. The bamboo sound listeners put their special feelings into the branches and leaves, which in the eco-ethics perspective today, contains some valuable spiritual elements that the contemporary people have been gradually missing.

Nevertheless, no matter what kind of satisfaction the listeners have derived from and what emotion they expressed through the behavior; "listening to bamboo" itself was a very individualistic act. It reflected to some extent the literati's psychological escape from a distressing reality. It is very hard to estimate the extent to which this behavior was influenced by Buddhism and Taoism, but from many related accounts we can feel their strong desire for the "Territory beyond" or "Taoyuan outside the world" (*shiwai taoyuan* 世外桃源). Most of the listeners only focused on their personal feelings and desired to return to a quiet, pure and comfortable life. They rarely paid attention to the social reality and the public weal and woe. Hence, "listening to bamboo" was only an elegant lifestyle of the rich people who had too much leisure, sometimes they were just moans and groans without being ill. However, Zheng Xie (鄭燮, i.e., Zheng Banqiao 鄭板橋, 1693–1766), a great painter in the Qing Dynasty, was an exception.

Among the poets and painters in ancient times, there was nobody who loved bamboo as deeply as Zheng Xie and nobody painted bamboo as uniquely as he did. About "listening to bamboo," he had some special feelings and understandings which are reflected in many of his works including poems and paintings. He very much appreciated the artistic conception of poetry and painting created by Wu Zhen (吳鎮, 1280–1354, also known as Mei Daoren 梅道人), a famous painter in the Yuan Dynasty, who once said: "I have a pavilion in the deep bamboo forest and like to listen to the sound of autumn in it." Zheng Xie applied the concept to his paintings and poems about the sound of bamboo quite a few times and his works were rich in aesthetic implications. It seemed that his ears were always full of the sounds of "bamboo wind" and "bamboo rain," and musings arose spontaneously in his mind whenever he heard sounds from bamboo, even if it was in deep night. In a poem he wrote: "I only need two poles of bamboo in front my window, so I can enjoy the excellent sound and the moisture of wind and rain. Being afraid to wake up others, I finish creating a poem without leaving the pillow."[37]

Although there were sounds of wind and rain in the bamboo grove throughout the year, what moved Zheng most were "the autumn sounds"; they were not necessarily real autumn sounds, but rather sounds in a bamboo grove that sounded like autumn sounds. It might not have rained, it might have been actually the wind shaking bamboo, blowing through the branches and leaves and producing noises that sounded like rain. It seemed that "the autumn sounds" had been mentioned in most of his bamboo poems. He was the person who pushed the image of the sound of bamboo to an extreme state. Generally, people need related vocabularies to describe "the sound," in a poem of his, however, there is not a single word directly referring to the sound or the noise, but it seems the reader's ears are full of bamboo sounds. He described that: "There are more autumn winds in the bamboo grove and the bamboo with whirling shadows beats my door and window. I refuse to remove them in order to prevent the Sandman from coming at midnight."[38]

Different from the common bamboo-sound listeners, Zheng was neither a dodger of reality, a hermit pursuing an elegant private life in the mountains abounding with forests and springs, nor a bureaucrat who did not care about the sufferings of the ordinary people. His heart was filled with care and sympathy. It was due to this care and sympathy that he said in a poem that he could hear the people's suffering in the rustle of bamboo in the cold autumn:

> Lying in the government house and listening to the rustling of bamboos outside, I suspect it is the poor people's lament. Though a small official of very low rank, each branch and leaf related to my concerns for the people related.

By using the word "yi 疑" (suspect), he expressed his deep concern for the sufferings of the people, and created a poetic masterpiece about bamboo sound of a highest standard through the ages. Just because of such a suspicion, he made all bamboo poets ashamed and numberless similar poems lost color! He deserves to be recognized as the best of the ancient bamboo lovers.

The bamboo grove habitat and the acoustic landscape created by the literati

Forestry experts pointed out: "Because of the reflection, absorption, and blocking effect to sound wave, bamboo forest or grove can greatly reduce noise level. Depending on a 40 meter wide bamboo forest belt, noise can be reduced by 10~15 dB."[39] This suggests that there are scientific grounds for the ancient literati keen on living in a bamboo grove. They chose to build residential and recreational sites in bamboo groves, or planted bamboo in gardens or around houses, so as to create a unique sound environment and soundscape. In fact, bamboo was an important material in ancient Chinese construction. "Bamboo buildings" once were very common south of the Qinling Mountain (秦嶺) and the Huaihe River (淮河), but now are strictly limited to a small region in southwest China. Bamboo also was one of the most important landscape plants. Since Wang Huizhi (王徽之, ?–388) and his like generated the ethos of loving bamboo in the period of the Eastern Jin Dynasty, planting bamboo in courtyards or around houses has become increasingly common. Poets such as Xie Tiao (謝朓, 464–499), Jiang Yan (江淹, 444–505), He Xun (何遜, 480–520), Shen Yue (沈約, 441–513), Jiang Hong (江洪, ?–?) and Zhang Zhengjian (張正見, ?–ca. 575) in the period of the Southern Dynasties occasionally mentioned in their poems bamboo groves (either wild or cultivated) nearby or around different constructions such as rooms, halls and pavilions, etc.

Up to the Tang Dynasty, as the cultural status of bamboo constantly rose, the literati intentionally planted bamboo as "partners" to create a living environment with the participation of bamboo. This kind of activity gradually prevailed in the literati society. For instance, the "Ten Scenic Sites in Wangchuan" (*Wangchuan shijing* 輞川十景) where Wang Wei, a leading poet of the mid-Tang dynasty, lived had very beautiful sceneries. Bamboo played an irreplaceable role in the sites including "Zhuli guan (竹裡館)" and "Jingzhu ling (斤竹嶺)." Wang frequently described the deep quiet and pleasant bamboo forests, praised the pure green, elegance and clarity of the bamboo which brought him a peaceful and quiet life away from the hubbub, and which served as a foil to his own elegance and dignity. No wander Wang's works were praised as "painting-in-poetry" (*shizhong youhua* 詩中有畫) and "poetry in a painting" (*huazhong youshi* 畫中有詩) and had a far-reaching impact on the poetry, painting and garden art of later periods.

Du Fu (杜甫, 712–770), another leading poet in the Tang period, once lived in Jiannan (劍南, in Sichuan province today) for quite a few years as an immigrant refugee from the An Lushan (安祿山) Rebellion (755–763). His humble cottage (Du Fu Caotang 杜甫草堂) was built with bamboo and surrounded by bamboo grove. He professed himself that "I must plant some bamboo wherever I live."[40] He summarized the common preference for bamboo of many literati in his era.

A more typical person was Bai Juyi. Although born in the northern area, he had a very rich life experience in southern China and liked to plant bamboo around his house. When he was a youth living in Changan (長安, today's Xi'an 西安) for the first time, he planted plenty of bamboo in his yard or around his houses. Later he traveled to and lived in the south for many years as a local official. The southern sceneries, especially the bamboo groves, had a great influence on his life. When he was old, Bai moved back to the north and lived in Luoyang (洛陽). Bamboo occupied one-ninth the space of his private garden which was roughly two acres, and 1,000 poles of bamboo were planted.[41] This showed his deep love for bamboo. Bai once said in a poem: "Water has a pure and clear character, so it is my friend; while bamboo is humble and modest, it is my teacher." Wherever he moved to and whatever situation he was in, he would plant some bamboo, not only because of his love for the green color and the slender figure, but also in order to listen to the sound as we mentioned above.[42]

The Tang Dynasty literati planted bamboo not only in their private gardens, but also around the government office buildings in order to construct a quiet and elegant living and working environment. It gradually became a strong cultural tradition of living, which continued to flourish until modern times. They planted bamboo for both visual and auditory environment entertainment. They actually took bamboo as neighbors or close friends as they lived side by side with it, day and night.[43]

In the Song Dynasty, bamboo was a common elegant guest in the private and public gardens. Zhou Weiquan (周維權, 1927–), a modern historian specialized in gardens, pointed out that: "The private gardens in Luoyang in the Song Dynasty were famous for flowers and trees. There were large tracts of grove landscape including bamboo, plum, peach, pine and cypress grove; and bamboo groves were especially common." He summarized the four major characteristics of the private gardens in the Song Dynasty as "simple" (*jianyuan* 簡遠), "bright" (*shulang* 疏朗), "elegant" (*youya* 優雅) and "natural" (*ziran* 自然). Bamboo was the embodiment of the "elegant." *The Record of Luoyang Gardens* [*Luoyang Ming Yuan Ji* 洛陽名園記] has recorded 19 of the city's gardens and most of them had a bamboo grove. It was said that water, bamboo and houses cover three-tenths, two-tenths and one-tenth in the total area of a garden respectively.[44]

Gardens in southern China even couldn't be constructed without bamboo. In the watery environment of Jiangnan, water and bamboo were in fact the two essential elements of a garden. Su Shunqing (蘇舜欽, 1008–1048), a Song Dynasty writer, once described the setting of his garden named "Canglang Ting 滄浪亭" as:

> A bamboo grove is in the front part, water (pond) in the back, and countless bamboo again in the north of the water (pond). The stream is clear and the bamboo green. Lights and shadows converge in doors and windows, and they constitute a scene that is harmonious with the wind and the moonlight.

Since then, "Shuizhuju 水竹居," a residence rich in water and bamboo, has become a common and typical eco-cultural landscape in the Jiangnan area.

The *Yuan Ye* [園冶] written by Ji Cheng (計成, 1582–1642), an architect and landscape-artist in the Ming Dynasty, is known as a classical book of landscape architecture. It catalogs dozens of scenes and discusses in detail how to use a variety of natural things to create different sceneries, among which 20 scenes are related to bamboo. This clearly illustrates the important role of bamboo in ancient garden landscaping, and fully explains why many places were called "Listening Bamboo" or similar names referring to bamboo.

In the Ming and Qing periods, many literati illustrated the living places in relation to bamboo and vividly described the acoustic landscapes in poems, essays and paintings. For example, Zheng Xie once described in a prefatory of his ink-bamboo painting as follows:

> A humble thatched cottage; several poles of fresh bamboo, the snow-white paper window slightly stained green color [because of the fresh bamboo]. At that time, I sit in alone, drinking a cup of Yuqian tea [雨前茶, i.e., the tea picked before the sixth solar term, *kuyu* 穀雨, around April 20], using a ink-stone from Duanzhou [Duanyan 端硯] and a piece of paper produced in Xuanzhou [Xuanzhi 宣紙], drawing a flower with several strokes. When a friend comes, the sounds of wind and bamboo just add to the quiet and

> silence. The houseboy cleans the room and the maidservant burns incense. They walk to and fro in the bamboo shadow and the light reflects on the painting. In such a scene of extreme beauty, why should I need to have 12 well-dressed girls or 100 actresses? What I need to do is to enjoy the refreshing breeze and the quiet sounds of the bamboo.[45]

Such quiet living circumstances similar to those described by Zheng can be found frequently in ancient Chinese documents, particularly in the literati's collections of late imperial times. There were also many paintings which illustrate this kind of living environment and help us to imagine how the sounds of the bamboo groves enhanced the silence. All kinds of related materials confirmed that the ancient literati constructed their bamboo habitats in pursuit of a quiet, elegant and leisurely life. The key concepts, "clear" (*qing* 清), "quiet" (*you* 幽), "elegant" (*ya* 雅) and "delight" (*qu* 趣), reflected not only the scholars' attitudes towards to life, but also embodied their natural aesthetic consciousness.

From abundant poems and paintings, we can obtain a unique and beautiful image of their ideal living environment as following: Clusters of bamboo growing luxuriantly around a humble hut, tall and handsome, with poles as smooth as cold jade, strong joints and bluish green leaves. Following the rhythm of nature, such as the changing of seasons and weather along with the sunlight in the morning, the sunset glow at dusk, the moonlight, the twinkling stars, the wind and rain, the frost and snow, the clouds and fog, the tree and grass, the birds and insects, the bamboo grove was constantly changing its colors, figures, flavors, shadows and sounds. All the sounds, no matter those of the blowing wind, the rustling rain, the singing birds, the hoarse cry of the autumn cicada, or the breaking of frozen bamboo poles and the elastic swinging of bamboo branches covered with heavy snow, were never perceived as noisy and annoying, but rather as supplements to the quietness and calmness. The residents pondered philosophical problems, recited poems, drew paintings, played the zithers and chess, and drank wine or tea in the bamboo grove. They lived leisurely, calmly, elegantly and otherworldly, and not only appreciated the grace, bearing and color of the bamboo, but also enjoyed its sounds. They faced "this gentleman" (*cijun* 此君), an anthropomorphic elegant name that the ancient Chinese literati offered to bamboo, all day long, and communicated with the bamboo spiritually until they reached a state of forgetting both the bamboo and themselves.

To people today, this is a mysterious realm just like the sound of footsteps in a deep valley, audible but impossible to chase.

Notes

1 Peter A. Coates, "The Strange Stillness of the Past: Toward an Environmental History of Sound and Noise," *Environmental History* 10.4 (Oct., 2005): 636–665.

2 Yuan Xiaomei 袁曉梅 et al., "Control of Environmental Sounds in Classical Chinese Gardens (中國古典園林的聲環境控制)," *Traditional Chinese Architecture and Gardens* [古建園林技術], 2006(3): 21–24. Yuan Xiaomei et al., "Construction of Soundscape in Classical Chinese Gardens (中國古典園林的聲景觀營造),"

Architectural Journal [建築學報], 2007(2): 70–72. Yuan Xiaomei, "The Formation and Evolvement of Soundscape Thought of Chinese Classical Garden (中國古典園林聲景思想的形成及演進)," *Chinese Landscape Architecture* [中國園林], 2009(7): 32–38. Wang Shuyan 王書豔, "Soundscape: The Sense of Sounds Regarding Gardens in Tang Poetries (聲音的風景: 園林視域中的唐詩聽覺意象)," *Social Sciences in Yunnan* [雲南社會科學], 2012(3): 144–148. Yang Jun 楊俊, "Landscape with Voiceless, the Soundscape Construction of Plant Arrangement in Jiangnan Classical Private Gardens (山水有清音—論江南古典私家園林植物配置的聲景營造)," *Art and Design* [藝術與設計理論], 2013(3): 87–89.

3 Wang Hongyuan 王宏遠, "A Random Discussion on Description of Sounds in Classic Poetries (古詩聲音描寫摭談)," *Journal of Language and Literature Studies* [語文學習], 2011(7): 56–57.

4 Wu Shuoxian 吳碩賢, "Soundscape Described in the Book of Poetry (詩經中的聲景觀)," *Architectural Journal* [建築學報], 2012(1): 109–113.

5 Jing Kaicheng 金開誠 et al., *Annotation of Quyuan's Works* [屈原集校注], Vol. 1, p. 280.

6 Yao Silian 姚士廉, *History of Liang Period* [梁書], Vol. 50, "Biography of Wang Ji (王籍傳)," p. 713.

7 Peng Dingqiu 彭定求 (ed.), *Complete Tang Poetry* [全唐詩], Vol. 128, p. 1302.

8 Shang Yongliang 尚永亮, "A Preliminary Study on the Sense of Autumn Feeling (悲秋意識初探)," *Journal of Shaanxi Normal University (Philosophy and Social Sciences Edition)* [陝西師範大學學報 (哲學社會科學版)], 1988(4): 31–38. Zhao Mingli 趙敏俐, "Autumn and the Theme of Yearning for Love and Native Place in Chinese Literature (秋與中國文學的相思懷歸母題)," *Social Sciences in China* [中國社會科學], 1990(4): 63–78.

9 Ouyang Xiu 歐陽修, "A Prose of Autumn Sound (秋聲賦)," *Collection of Ouyang Xiu's Works* [歐陽文忠公集•居士集], Vol. 15.

10 Ouyang Xun 歐陽詢 et al., *Collection of Literature Arranged by Categories* [藝文類聚], Vol. 89, p. 1552.

11 *Complete Tang Poetry*, Vol. 334, p. 3747.

14 Ibid., Vol. 432, p. 4807; Vol. 449, p. 5059.

15 Ibid., Vol. 433, p. 4792.

16 Ibid., Vol. 560, p. 6502.

17 Ibid., Vol. 574, p. 6690.

18 Ibid., Vol. 707, p. 8135.

19 Ibid., Vol. 672, p. 7686.

20 Ibid., Vol. 708, p. 8153. "Six Tonalities" (六律) refers to Huangzhong (黃鐘, C), Taicu (太蔟, D), Guxi (姑洗, E), Ruíbin (蕤賓, #F), Yíze (夷則, #G) and Wuyì (無射, #A).

21 *Complete Tang Poetry*, Vol. 766, pp. 8695, 8689, 8698.

22 Ibid., Vol. 746, p. 8489.

23 Sun Di 孫覿, "Stay in Jiusi-An in Snowy Night (雪夜宿九思庵)," *Collection of Hongqing Jushi* [鴻慶居士集], Vol. 6.

24 Ma Zhi 馬治 and Zhou Di 周砥, "Sitting by the West Pond and Listening to the Wind-blowing Bamboo in Night (西池夜坐聽風竹)," *A Collection of Poems in Jingnan* [荊南倡和集].

25 Shen Zhou 沈周, "Listening to Bamboo (聽竹)," in *Selected Poems of Shen Zhou* [石田詩選].

26 Zhao Mengfu 趙孟頫, *Collection of Proses from Song Xue Zhai* [松雪齋文集].

27 Wang Anshi 王安石, "Six Miscellaneous Poems (雜詠六首)," in Li Bi 李壁, *Annotation of Poetry of Wang Jinggong* [王荊公詩注], Vol. 46.

28 Lu You 陸遊, "Sitting and Listening to the Rain Sound in the Bamboo Grove (夜坐聽竹間雨聲)," adapted from Qian Zhonglian 錢仲聯, *Annotation of Poems of Lu You* [劍南詩稿校注], Vol. 17, p. 1317.

29 Wu Yulun 吳玉綸, "A Note of the Picture of Listening to Rain by the Bamboo Window (竹窗聽雨圖記)," *Drafts of Xiangting* [香亭文稿], Vol. 5.
30 Song Ne 宋訥, "Ge Xing 歌行 (a style of ancient Chinese poems)," *Collection of Xiying* [西隱集], Vol. 4.
31 Qian Qianyi 錢謙益, "Stay in Ganlu'an (宿甘露庵)," *Collection of Poetry in Reigns of Ming Dynasty, The Run Collection* [列朝詩集•閏集], Vol. 3.
32 Shen Shourong 沈壽榕, "Two Poems to Zhao Binru (趙冰如見過卻贈二首)," *Collection of Poetry from Yushenglou* [玉笙樓詩錄], Vol. 2.
33 In Fa Shishan 法式善, *Poems of Wumeng* [梧門詩話], Vol. 1.
34 Zhang Yingchang 張應昌, "A Poem of Plum Blossom (梅花引)," *Fisherman's Song in Mists and Rippers* [煙波漁唱], Vol. 2.
35 Wen Zhengming 文徵明, "A Picture of Listening to Jade (聽玉圖)," in Wang Keyu 汪砢玉, *A Collection of Calligraphy and Painting* [珊瑚網], Vol. 39.
36 Apart from Lu You, there were quite a few poets who had mentioned the effects of "listening to bamboo" in waking consciousness, eliminating anxiety and curing mental diseases.
37 Cao Huiming 曹惠民 et al. (eds.), *A Complete Collection of Zheng Panqiao's Poems, Proses, Calligraphies, and Paintings* [鄭板橋詩文書畫全集], p. 81.
38 Zheng Xie 鄭燮, *Collection of Zheng Banqiao's Works* [鄭板橋集], p. 209.
39 *Cai Baozhen* 蔡寶珍 et al., "Biological Characteristics of Bamboo and Its Application in Scenic Gardening," *World Bamboo and Rattan* [世界竹藤通訊], 2011(4): 39–43.
40 *Complete Tang Poetry*, Vol. 221, p. 2332.
41 Ibid., Vol. 461, pp. 5249–5250.
42 Ibid., Vol. 434, p. 4807; Vol. 432, p. 4768.
43 Qin Guan 秦觀, "Using the Rhyme Sequence of Zeng Cengzhi's *Poem of Xiao-zhu-xuan* (次韻曾存之嘯竹軒)," *Collection of Huaihai* [淮海集], Vol. 6.
44 Zhou Weiquan 周維權, *History of Chinese Classical Gardens* [中國古典園林史], Beijing: Tsinghua University Press, 1990, pp. 140, 145, 147.
45 Cao Huiming 曹惠民 et al. (eds.), *A Complete Collection of Zheng Panqiao's Poems, Proses, Calligraphies, and Paintings* [鄭板橋詩文書畫全集], p. 80.

Bibliography

Cai Baozhen 蔡寶珍 et al. "Zhuzi de shengwuxue texin jiqi zai fenfjing yuanlin zhong de yingyong 竹子的生物學特性及其在風景園林中的應用 (Biological Characteristics of Bamboo and Its Application in Scenic Gardening)." *Shijie zhuteng tongxun* 世界竹藤通訊 [*World Bamboo and Rattan*], 2011(4): 39–43.

Cao Huiming 曹惠民 et al. (eds.). *Zheng Banqiao Shi Wen Shu Hua Quan Ji* 鄭板橋詩文書畫全集 [*A Complete Collection of Zheng Panqiao's Poems, Proses, Calligraphies, and Paintings*]. Beijing: Zhongguo Yanshi Chubanshe 中國言實出版社, 2006.

Coates, Peter A. "The Strange Stillness of the Past: Toward an Environmental History of Sound and Noise." *Environmental History*, 10.4 (Oct., 2005): 636–665.

Fa Shishan 法式善. *Wumeng Shihua* 梧門詩話 [*Poems of Wumeng*]. A manuscript stored by Beijing Library 北京圖書館藏手稿.

Jing Kaicheng 金開誠 et al. *Quyuan ji Jiaozhu* 屈原集校注 [*Annotation of Quyuan's Works*]. Beijing: Zhonghua Shuju 中華書局, 1996.

Li Bi 李壁. *Wang Jinggong Shi Zhu* 王荊公詩注 [*Annotation of Poetry of Wang Jinggong*], in *Siku Quanshu* 四庫全書 [*Complete Library in Four Sections*].

Ouyang Xiu 歐陽修. *Ouyang Wenzhonggong Ji, Jushi Ji* 歐陽文忠公集·居士集 [*Collection of Ouyang Xiu's Works*], in *Sibu Congkan* 四部叢刊 [*Collection of Four Branches of Classics*].

Ouyang Xun 歐陽詢 et al. *Yiwen Leiju* 藝文類聚 [*Collection of Literature Arranged by Categories*]. Shanghai Guji Chubanshe 上海古籍出版社, 1965.

Peng Dingqiu 彭定求 (ed.). *Quan Tang Shi* 全唐詩 [*Complete Tang Poetry*]. Beijing: Zhonghua Shuju 中華書局, 1960.

Qian Qianyi 錢謙益. *Lie Zhao Shiji, Run Ji* 列朝詩集•閏集 [*Collection of Poetry in Reigns of Ming Dynasty, The Run Collection*]. Printed by Ji Gu Ge 汲古閣, the 9th year of Emperor Shunzhi 順治 (1652).

Qian Zhonglian 錢仲聯 (annotated). *Jiannan Shigao Jizhu* 劍南詩稿校注 [*Annotation of Poems of Lu You*], Shanghai: Shanghai Guji Chubanshe, 1985.

Qin Guan 秦觀. *Huaihai Ji* 淮海集 [*Collection of Huaihai*], in Siku Qianshu 四庫全書.

Shang Yongliang 尚永亮. "Beiqiu yishi chutan 悲秋意識初探 (A Preliminary Study on the Sense of Autumn Feeling)." *Shaanxis Shifan Daxue Zuebao* 陝西師範大學學報 [*Journal of Shaanxi Normal University* (*Philosophy and Social Sciences Edition*)], 1988(4): 31–38.

Shen Zhou 沈周, *Shitian Shi Xuan* 石田詩選 [*Selected Poems of Shen Zhou*], in *Siku Quanshu* 四庫全書.

Sheng Shourong 沈壽榕. *Yushenglou Shilu* 玉笙樓詩錄 [*Collection of Poetry from Yushenglou*], printed in the ninth year of Emperor Guangxu 光緒 (1883).

Song Ne 宋訥. *Xi Ying Ji* 西隱集 [*Collection of Song Ne*], in *Siku Quanshu* 四庫全書.

Sun Di 孫覿. "Xueye su jiuxian," in *Hongqing Jushi Ji* 鴻慶居士集 [*Collection of Hongqing Jushi*], in *Siku Quanshu* 四庫全書.

Wang Hongyuan 王宏遠. "Gushi shengyin miaoxie zhitan 古詩聲音描寫摭談 (A Random Discussion on Description of Sounds in Classic Poetries)." *Yuwen Xuekan* 語文學刊 [*Journal of Language and Literature Studies*], 2011(7): 56–57.

Wang Keyu 汪砢玉. *Shanhu wang* 珊瑚網 [*A Collection of Calligraphy and Painting*], in *Siku Quanshu* 四庫全書.

Wang Shuyan 王書豔. "Shengyin de fengjing: yuanlin shiyu zhong de Tangshi tingjue yixiang 聲音的風景: 園林視域中的唐詩聽覺意象 (Soundscape: The Sense of Sounds Regarding Gardens in Tang Poetries)." *Yunnan Shehui Kexue* 雲南社會科學 [*Social Sciences in Yunnan*], 2012(3): 144–148.

Wu Shuoxian 吳碩賢. "Shijing zhong de shengjingquan 詩經中的聲景觀 (Soundscape Described in the Book of Poetry)." *Jianzhu Xuebao* 建築學報 [*Architectural Journal*], S1 (2012): 109–113.

Wu Yulun 吳玉綸. *Xiangting Wengao* 香亭文稿 [*Drafts of Xiangting*]. Printed by Zi-De-Tang 滋德堂 in the 60th year of Emperor Qianglong 乾隆 (1795).

Yao Silian 姚士廉. *Liangshu* 梁書 [*History of Liang Period*]. Beijing: Zhonghua Shuju 中華書局, 1973.

Yang Jun 楊俊. "Shanshui you qingyin—lun Jiangnan gudian sijia yuanlin zhiwu peizhi de shengjing yinzao 山水有清音—論江南古典私家園林植物配置的聲景營造 (Landscape with Voiceless, the Soundscape Construction of Plant Arrangement in Jiangnan Classical Private Gardens)." *Yishu yu sheji lilun* 藝術與設計理論 [*Art and Design*], 2013(3): 87–89.

Yuan Xiaomei 袁曉梅 et al. "Zhongguo gudian yuanlin de sheng huanjing kongzhi 中國古典園林的聲環境控制 (Control of Environmental Sounds in Classical Chinese Gardens." *Gujian yuanlin jishu* 古建園林技術 [*Traditional Chinese Architecture and Gardens*], 2006(3): 21–24.

Yuan Xiaomei 袁曉梅 et al. "Zhongguo gudian yuanlin de sheng huanjing yingzao 中國古典園林的聲景觀營造 (Construction of Soundscape in Classical Chinese Gardens." *Jianzhu xuebao* 建築學報 [*Architectural Journal*], 2007(2): 70–72.

Yuan Xiaomei 袁曉梅. "Zhongguo gudian yuanlin shengjing sixiande xingcheng ji yianbian 中國古典園林聲景思想的形成及演進 (The Formation and Evolvement of Soundscape Thought of Chinese Classical Garden)." *Zhongguo yuanlin* 中國園林 [*Chinese Landscape Architecture*], 2009(7): 32–38.

Zhang Yingchang 張應昌. *Yanbo Yuchang* 煙波漁唱 [*Fisherman's Song in Mists and Rippers*], printed by Xichang Lushe 西昌旅舍 in the second year of Emperor Tongzhi 同治 (1863).

Zhao Mengfu 趙孟頫. *Song Xue Zhai Wenji* 松雪齋文集 [*Collection of Proses from Song Xue Zhai*], in *Siku Quanshu* 四庫全書.

Zhao Mingli 趙敏俐. "Qiu yu zhongguo wenxue de xiangsi huaigui muti 秋與中國文學的相思懷歸母題 (Autumn and the Theme of Yearning for Love and Native Place in Chinese Literature)." *Zhongguo Shehui Kexue* 中國社會科學 [*Social Sciences in China*], 1990(4): 63–78.

Zheng Xie 鄭燮. *Zheng Banqiao Ji* 鄭板橋集 [*Collection of Zheng Banqiao's Works*]. Beijing: Zhonghua Shuju 中華書局, 1962.

Zhou Di 周砥 and Ma Zhi 馬治. *Jingnan Changhe Ji* 荊南倡和集 [*A Collection of Poems in Jingnan*], in *Siku Quanshu* 四庫全書.

Zhou Weiquan 周維權. *Zhongguo Gudian Yuanlin Shi* 中國古典園林史 [*History of Chinese Classical Gardens*]. Beijing: Tsinghua University Press 清華大學出版社, 1990.

2 Livestock and the "industrious revolution" in Tokugawa Japan (1603–1868)

Satoshi Murayama

Introduction: two types of "industriousness" in early modern times

This chapter will examine two kinds of "industriousness" and their differing viewpoints,[1] which are characterized in the four passages provided below. The first description by the Duke of Württemberg (Frederick I, 1557–1608) shows the essence of the freedom of consumers, especially women. The second excerpt from Sir James Steuart (1712–1790)'s writings explains how "industriousness" is derived from the compulsion of market. The subsequent two paragraphs from Japanese farm manuals, in contrast, present the idea that farm property belongs not to the individual but rather to the family, and should be passed from ancestors to descendants and that the purpose of hard work was to escape poverty by pursuing the disciplined use of the time necessary for "industriousness."

> The women have much more liberty than perhaps in any other place; they also know well how to make use of it, for they go out dressed in exceedingly fine clothes ... [England is] a paradise for women, a prison for servants, and a hell or purgatory for horses ... for the females have great liberty and are almost like masters, whilst the poor horses are worked very hard.
>
> (The Duke of Württemberg on his 1598 visit to England[2])

> In the ancient world, "men were ... forced to labor because they were slaves to others; men are now forced to labor because they are slaves to their own wants." As a result, "in a trading nation every man must turn his talents to account, or he will undoubtedly be left behind in this universal emulation, in which the most industrious, the most ingenious, and the most frugal will constantly carry off the prize."
>
> (Sir James Steuart, *Inquiry into the Principles of Political Economy*, 1767[3])

> The farm family consists of the fields, wealth, and heirlooms handed down from ancestors. This property does not belong to us, the living members of the family. We must not imagine it does even in our dreams. It belongs to the ancestors who founded the house; we are only entrusted with its care and must pass it on to our descendants.... There may be events beyond our

> control, such as flood, fire, or illness, as a result of which the sale of property becomes unavoidable. In that case, we must make every effort by saving and planning to recover what has been sold, make the property whole again, and pass it on undiminished to our children and grandchildren.
>
> (*Nogyo Yoshu* 農業要集, a farm manual, 1826[4])

> If the farm family would escape poverty, it must treat time as precious.[5] By rising early and shortening the daily rest period, two additional hours a day can be worked. That is seven hundred and twenty hours a year: the equivalent of sixty days, or two months, when no food is consumed, no wage paid, no oil required for lighting.... Thus can the farm family escape the pain of poverty, raise itself up, illuminate the deeds of ancestors, and confer blessings on descendants.
>
> (*Nogyo Mokun* 農業蒙訓, a farm manual, 1840[6])

Hayami Akira contrasted the labor-intensive technologies of Tokugawa Japan (1603–1868), in early modern Japan, with the capital-intensive technologies of Britain during the Industrial Revolution.[7] According to Saito Osamu, Hayami's original "industrious revolution" theory consisted of four parts.[8] (1) Between the sixteenth and seventeenth centuries, Japan experienced transition from a command economy to a market economy. (2) This "commercialization" prompted an increase in population and a concomitant decrease in the land-population rate. (3) The decrease in the amount of land available changed the typical household size to smaller units. Complex kinship family farms with subordinate laborers were reduced to simple family farms, where the labor force was limited to members of the immediate family. (4) In the process, Hayami used the term "industrious revolution" to account for "the growth in agricultural output in the last decades of the Tokugawa era that was achieved as peasants adapted their farming methods to substitute increased human exertion for the tractive power of livestock."[9] As Hayami himself explained, "there must have been a conversion from 'horse power' to 'manpower' in rural Japan; and the term 'Industrious Revolution' can be applied to describe this change."[10]

Two theories of the "industrious revolution"

Under the political initiative of Toyotomi Hideyoshi (豐臣秀吉, 1537–1598), a nationwide survey of arable lands began in 1582. The land survey, which investigated the whole of Japan from the end of the sixteenth to the beginning of the seventeenth century in the Tokugawa era, stimulated historical debate in Japan beginning in the mid-1950s: Under the influences of the terms and theories of Marxism, economic historians claimed that the land survey caused an institutional revolution from a "slave" system to a "feudal" society occurred by establishing "independent peasants" in Japan. It is now still questionable whether the changes in agricultural land and life were really that drastic, because Hayami, according to his historical-demographic village studies,[11] characterized the historical change not as a social revolution based on an institutional transition but as a demographic transition caused by the changing processes of household and family labor management.

The household composition changed from "family," which included not only kinships, but also laborers and servants, to "family" consisting only of intimate kinship relations in the modern sense.[12] The timing and period of the transition of family system was diversified regionally, but only in terms of the increase in peasant "family" farming. De Vries revised this concept to adapt the argument to be more suitable to early modern commercialization in Europe. He insisted that a fundamental change occurred in the household economy, thus increasing the labor supply for the market.[13] The European redefinition of the "industrious revolution" has proponents who provide empirical evidence of the revolution. Households increased their labor market participation in order to buy new consumer goods with the money earned.[14] In this European pattern of the "industrious revolution," the expansion of work hours correlated with the expansion of an economy of consumption. De Vries demonstrated "that historical consumption has been a dynamic phenomenon, charting a far from linear process of change."[15]

De Vries's revision opened up a new dimension of debate concerning economic development from pre-modern to modern times. Sugihara Kaoru, Saito Osamu, and other scholars expanded on his argument, and hence the term "industrious revolution" came to mean the "general process of labor-intensive industrialization, which contributed to the non-Western countries to achieve modern economic growth."[16]

Hayami's views were reduced to a simple path towards industrialization. The "industrious revolution" came to be regarded as "a process of household-based resource reallocations that increased both the supply of marketed commodities and labor and the demand for market-supplied goods."[17] We must remember that Hayami never discussed the transition from the "industrious" to the "industrial" revolution, nor did he focus on an immense labor supply in relation with the increase of productivity because Japanese "ecological conditions" were more suitable to small family farming, which could bring much more efficiency than a big farm using paid labor.[18]

Both Japan and Western Europe experienced an increase in industrial and commercial activity in early modern times, but there was a crucial difference in their urbanization processes. According to Thomas C. Smith, pre-modern growth in the West was stimulated by urban growth; in contrast, Tokugawa Japan lost population in cities and towns in the latter half of the era.[19] De Vries modified this theory of the "urbanization" process in European countries:

> While the population remained overwhelmingly rural until the nineteenth century it did not remain overwhelmingly agricultural, at least not if employment is defined by the actual allocation of days of labor rather than by the persistence of a minimal attachment to the land. As the rural population came to be organized into more effective market networks, it acquired a more complex occupation character.[20]

This is one of the main arguments by which de Vries was able to characterize the "industrious revolution" within a global perspective. The crucial difference between Japan and the West was the changing attitude of rural populations who had to seek non-agricultural paid work in rural regions. In the Japanese case, there was no increase in a proletariat composed of rural, paid, landless laborers.[21] Hayami insisted that with the development of individual family farming, all family members had to work harder to gain more yields. All rural people in early modern Japanese villages belonged to existing families, without exception. Almost all case studies of early modern Japanese population registers, such as the *Shumon Aratame Cho* [宗門改帳], prove that it was extremely rare for an individual to live alone in a village.

Hayami's original idea was fundamentally different from de Vries's, because whereas Hayami's "industrious revolution" was a counterpart of the Industrial Revolution, de Vries regarded the industrious revolution only as a preparatory phase of the subsequent Industrial Revolution. The first half of Tokugawa Japan was a successor of the Japanese medieval world, but also a period of expansion in which wet paddy fields spread all over the country, bringing about fundamental change in the Japanese landscape.

Because Hayami's original concept of the "industrious revolution" focused on human labor as an alternative to livestock labor, his arguments can be reasonably extended to environmental issues regarding the relationship between humans and animals. Needless to say, pre-modern societies were not simply based on natural and ecological foundations. Interactions between humans and natural environments have a complex history. The point is that at no point in the debate regarding the "industrious revolution" has such an argument been raised.

Early modern development of market goods in Europe can be analyzed using probate inventories having their own history, which indicate that the proliferation of market goods had occurred prior to the Industrial Revolution, especially in England and the Netherlands. Probate inventories have historical significance as such recordings were made for a reason. Probate inventories were products of the early modern European world, which preceded the escape from agricultural constraints.[22] However, such significant sources for the analysis of the commercialization process are completely missing in Japanese early modern history.

The property of individuals and families were never officially registered in Tokugawa Japan, even though the economic growth of agrarian productivity was comparable between Japan and the West, as T. C. Smith and Saito clearly argue. Scheilagh Ogilvie analyzed the influence of social capital in early modern Germany, finding it affected the work and consumption behavior of rural inhabitants, especially of women.[23] In the Japanese case, a social institution such as the *shoya* (庄屋) village system had initial effects on early modern commercialization, because not the individual household but the village was collectively a unit for tax payment in Tokugawa Japan.

The output of each village in Tokugawa Japan was calculated in *kokudaka* (石高), a tax measurement equivalent to the quantity of rice the village was capable of producing. This measurement system was instituted in 1598. National aggregates exist for the years 1598, 1645, 1697, 1721, 1830, and 1872. Under the Tokugawa system, every piece of taxable land was assessed in terms of its productive capacity expressed in *koku* (石, 1 *koku*=150 kilograms of rice volume or capacity), which represented a quantity of rice. The entire farm output, covering all agricultural products, would be converted to *kokudaka*, and because the *kokudaka* served as a basis for determining the tax amount payable by a village, it is believed that from the 1697 compilation onwards, the gap between the official assessment and actual production widened due to the growth in productivity that undoubtedly took place from the mid-seventeenth century. In the Tokugawa period, it was widely held that 1 *koku* was enough to feed one person for one year. According to the calculations of Nakamura, the production of farm products in the benchmark year 1700 was 169 kilograms, exceeding the 150 kilograms (or 1 *koku*) level per person. By 1872, actual productivity had increased to 201 kilograms.[24]

A *shoya* was a peasant in an early modern village in Japan; however, *shoya* is also the name for the administrative representative, the headman of the village. In most villages in Japan, the village head and administrator, i.e., the *shoya*, managed the village and kept an administrative diary for the continuity and security of the local governance. Some diaries are available for analysis, and those of Takahama (高濱), a village in the Amakusa (天草) Islands, Kyushu (九州), Japan, are the primary sources for this paper.[25] Almost all local historical documents, including population registers, correspondences, official letters, and village budgets from the eighteenth and nineteenth centuries were safely stored away and are available today. The position of the *shoya* was mostly hereditary, and thus rural historical materials were kept in the house of the *shoya*'s family. The village in Tokugawa Japan as a unit of tax payment often functioned as a unit of solidarity and responsibility for sustainable development.

Population change and the expansion of wet rice fields

The Japanese population was about five million in 725, and increased slowly to 12 million at the beginning of Tokugawa Japan (see Figure 2.1). Japan was then isolated from the world for more than two and a half centuries due to the Tokugawa *sakoku* (鎖国, national isolation) policy. From the sixteenth to the eighteenth century, the population increased rapidly to 31 million, and then stagnated at around 32 million until the beginning of the Meiji (明治) Era from 1868.[26]

At the beginning of the twentieth century, the Japanese population was about 44 million; by 1950 it had surpassed 83 million, and by 2000, it reached 127 million. In other words, the Japanese population doubled in the first half of the twentieth century and tripled over the 100-year period. A 2005 national census found that Japan's population totaled slightly more than 127.75 million, down 19,000 from the year before, which means that the population actually peaked in 2004, two years earlier than had been predicted. Japan has witnessed a distinctive turning point in its demographic history—from a century of increase in the twentieth century to the twenty-first century, characterized by a decreasing

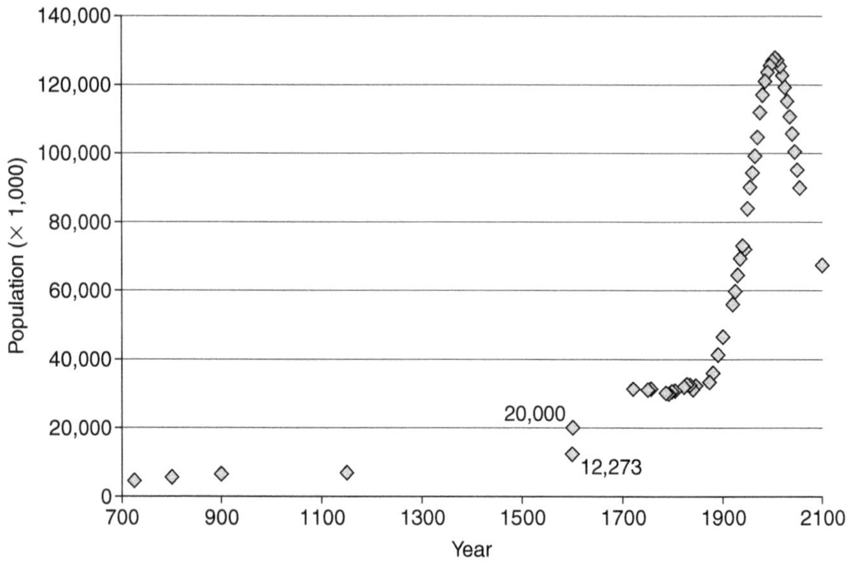

Figure 2.1 Historical change of Japanese population: two patterns of estimation (source: Kitou Hiroshi 鬼頭宏, *Japanese History Viewed from the Population* [人口から読む日本の歴史], (Tokyo: Kodansha, 2000); Moriyuki Oe, "Problems and Implications of Japan's Aging Society for Future Urban Developments," *Policy and Governance Working Paper Series*, No. 89 (March 2006); Tomobe Kenichi 友部謙一, *Peasant Household Economies in Pre-industrial Japan: Subjective Equilibrium and Market Economy* [前工業化期日本の農家経済—主体均衡と市場経済—], (Tokyo: Yuhi-kaku, 2007), 37–42).

population.[27] The population is expected to fall to 100 million by 2050 and to 64 million by 2100. The fertility problem is bound to change the country.

Drastic population increase occurred two times in Japan, between the sixteenth and seventeenth centuries, and in the latter half of the nineteenth century, especially after the Meiji Restoration following the Industrial Revolution in Japan. Hayami estimated the population of Japan at the beginning of the seventeenth century to be between 9.8 and 12.0 million; that number then jumped by more than 2.3 times to about 26.0 million. The latter number was calculated from the first national population survey, which took place in 1720. The former number, for the beginning of the sixteenth century, is an estimate, and is different from another estimate that put the number simply as "less than 20.0 million."[28]

The incredible expansion in the Japanese population up to the beginning of the eighteenth century was possible in part because agricultural output was able to keep pace, ultimately sustaining a population of 26.0 million people. The population increase was accompanied by a truly dramatic change of landscape in Japan during this period. For example, Mizumoto Kunihiko introduced an impressive change of landscape in Hineno (日根野), a village in the Kinki (近畿) region where the wilderness completely disappeared during this period. The wilds were transformed into wet paddy fields that covered the entire area of the village. Two maps for the same area are available for the years 1316 (Showa 正和5) and 1761 (Horeki 寶暦11). This represents a really rare case, in which we can compare the landscape changes in a 450-year period.[29]

The earlier pictorial map called "Hine-sho Hineno-mura Arano Ezu (日根荘日根野村荒野絵図)" (i.e., "a pictorial map of the wilds") was drawn in 1316 because plans for the expansion of the wet rice fields caused conflict between a landowner in the village and the village administrator. The development plan was not successfully executed until the second half of the seventeenth century, when the village inhabitants decided to perform excavation work.[30] They were able to join ponds and establish an irrigation system for a whole network of villages, including Hineno, Sano (佐野), Nagataki (長滝), and Kamisato (上郷). Rivers and ponds were joined both naturally and artificially. The water flows shifted the traditional village landscape from one dependent on rainwater to one reliant on an irrigation system that provided water incessantly, regardless of changes in seasonal rainfall.

The new irrigation system changed the landscape of the village Hineno quite clearly. The latter pictorial map called "Hineno-mura Igawa Yousui-zu (日根野村井川用水図)" (i.e., "a pictorial map for Igawa irrigation system in Hineno"), was drawn because of a conflict that occurred between the villages of Hineno and Sano in 1761 (Horeki 11). The irrigation system belonged not to individual peasants, but to the villages. The wilds had disappeared from the Hineno village. Wet rice fields covered the whole village.

The seventeenth century witnessed a huge expansion of wet rice fields in Japan. The new arable lands in the Kaga (加賀) countryside in the Tokugawa area, for example, were 40 percent established by 1645 (Shohou 正保2) and around 78 percent by 1683 (Tenna 天和3).[31] The Kanazawa (金澤) plain, the main living space in the Kaga region, was covered with a network of small waterways and brooks chained with ponds.

Mizumoto Kunihiko argued that this new landscape development had its roots in a "social revolution" that took place from the end of the sixteenth to the seventeenth century. He was speaking not of an economic stage of development in the traditional Marxist understanding of transition, but of a new dimension of revolutionary understanding concerning "callings stratification."[32]

In the late sixteenth century, warriors were scattered over the land in villages. They served as "overlords, levying taxes, administrating justice, and keeping the peace." "The normal state" among regional military banded organizations "was war or preparation for war." War was "the most direct means of increasing territory and thereby increasing strength and security."[33] The talented leadership of Nobunaga (信長), Hideyoshi (秀吉), and finally Ieyasu, Tokugawa Ieyasu (徳川家康, 1542–1616), unified Japan. Ieyasu destroyed the feudal leagues after a series of institutional changes under the regimes of the three leaders; the separation of warriors and farmers, a simple measurement system for land productivity, and isolation and anti-Christian politics brought a new but compromised political structure that managed to keep the peace for two and a half centuries. The new social norms denied the value of individual armed struggles by warriors to protect their own rights and properties. Instead, the legitimacy of a stratified system of callings was instituted. Professionalized business administration belonged to families that were admired and protected as a result. Kawachiya Kasho (河内屋可正, 1636–1713) described the norm succinctly:

> There are professional way of warriors, that of farmers, that of craftsmen, that of merchants, and also many other professional ways. If the people work hard along the way of the professionalized "family," the "family" will be richer, because the way must become main sources for gold and silver.[34]

Kasho himself had considerable status as a village headman in a suburb near Osaka, an economic center of Tokugawa Japan. He described a social order that brought money to individual families as professionalized households. Another state administrator described additional points of the state policy in the era of Kansei (寛永, 1624–1644):

> First, the peasant family is the basis of the world; second, the peasant should be managed so as not to possess too much, nor to live in scarcity; third, the government should pay to smooth the routes in September and October in order to improve traffic, and never use the people for other purposes.

This moral reasoning in politics was based on the belief and trust in the natural laws of heaven. Only a government following the natural laws of heaven will increase the wealth of the world under heaven.[35]

Okura Nagatsune (大藏永常, 1786–1856?) is one of the representatives of a large class of Tokugawa writers who may be called technologists.[36] He was a systematic observer: "wherever he went, he asked skilled farmers about local soils, seeds, fertilizers, irrigation, and tools, writing down what they had to say and making detailed sketches of plants, tools, and operations."[37] His first book concerning sericulture is titled *Farm Family Profits* [*Noka-eki* 農家益]. In the Tokugawa era there were many agricultural inventions that were "the result of centuries of adaptation of farming—of tools, seed, crops, tillage—to local conditions of climate and soil." For example, although "the spade was used everywhere … its size, design, and heft differed almost from village to village."[38]

Nagatsune and other technologists attached enormous importance to technical proficiency in farming. However,

> little mention was made of frugality or industry in the farmer, qualities that orthodox Confucian writers harped on [to the neglect of technical skill], clearly not because these virtues were thought unimportant by the technologists but because they were taken for granted.[39]

They also, however, "vehemently attacked religion and the idea of fate in its effect on farming," because "results were determined by skill, they insisted, not luck or divine help, and they insisted no doubt because contrary beliefs were widespread."[40] T. C. Smith introduced passages from a book on sericulture by an egg-grower in Shinshu (信州):

> Even in good growing years when everyone prospers, yields vary with skill. Although everyone is the same distance from Heaven, it is plain that there are differences in ability (*jinriki* 人力). People who do not recognize this fact stupidly pray to Buddha and *kami* (神), or they blame the eggs for their bad results and envy the success of others.[41]

"Industriousness" was taken for granted among professionalized farmers, and the accumulated technology was especially admired. But the wet rice fields were not isolated from ecological conditions of surrounding mountains and forestry. Their expansion changed also the surrounding mountains to grass fields for the sake of gaining access to compost.[42] The forest was lost to a large extent in the seventeenth century.[43] The increasing population stimulated market expansion for firewood, coal, and timber, which was another cause of deforestation. However, the crucial point is that the scattered rice field culture fostered interactive learning on the overall natural environment in Japan. Forestry also came to be characterized by a more regenerative concept. Fundamental ecological changes began from the expansion of wet rice fields, because the impacts upon the whole life community in the village and its surroundings were decisive for agricultural and social development in the following centuries.

Regional diversity in livestock patterns

The peasant family farm and the expansion of new arable lands in the seventeenth century increased the productivity of rice, but the overconcentration on a single crop increased the risk of famine. The first big famine in early modern Japan broke out in 1732 (Kyoho 享保17) mainly in western parts of the country. It developed into a serious famine caused by scarcity of rice and wheat, and the *Kokudaka* system, which was the foundation of the Tokugawa regime, failed spectacularly. For example, the land tax delivery from the Ozu han (大洲藩), which had remained stable at an average of 38,000 *koku* from 1727 to 1731 (Kyoho 12–16), dropped to 13,000 *koku* in 1732 (Kyoho 17), or about one third. The damage was serious. The severity of the famine prompted a big conversion in agriculture and industry from monoculture to polyculture production. Many domains in Tokugawa Japan began investigating the products available from the mid-eighteenth century. The listed diversification of local products provides evidence of the development of a new polyculture system.[44]

In Tokugawa Japan, epidemics and famines were the major causes of death; smallpox in particular was a key cause. Although the word "famine" does not appear in the diaries of Takahama,[45] "bad harvest" (*kyo-saku* 凶作 or *fu-saku*

不作) appears every three years or so, six times over 20 years in the available diaries. In serious cases of "bad harvest," the *shoya* tried to get local governments to reduce taxes. The early modern village people generally did not live in major flood-prone areas. Minor flooding was observed as normal events that did not lead to the deaths of humans or animals.

Facing the sea, Takahama did not have sufficient arable land for paddy fields that would collectively produce enough to meet the nutritional needs of the villagers. Takahama had a population of 3,413 in 1816 (Bunka 文化13), but the *kokudaka* of the village was only 611 *koku*, or 0.18 *koku* per capita. Takahama could not survive as an agricultural village.[46] Its subsistence was guaranteed only by the development of a complex commercialized economy based on sea transportation and export of ceramic stones.

Takahama's village economy changed remarkably from the end of the eighteenth to the nineteenth century. Two large commercial ships (nine-sails and six-sails) were introduced, together with an increasing number of small commercial boats, and these supported the commercialized village economy. Fishing and marine production increased in these decades, and in comparison with surroundings villages, Takahama came to be an especially commercialized village (Table 2.1). The number of fishing and seaweed boats increased from 12 in 1789 to 58 in 1838. There was also a decisive increase in other forms of transportation; for example, the number of horses increased from 117 in 1789 to 758 in 1838 (Table 2.2).

Such a marked increase in the means of transportation was likely one of the results of the commercialization of Takahama. Ceramic stones were mined in the mountain areas of Takahama, where high-quality stone for ceramic production

Table 2.1 Commercial and fishery tax (in "momme 匁" = mass unit for silver coin) on fishing- and transportation-related commerce in villages of the Oe district

Village	*Year*			
	1789		*1867*	
	Commercial tax	*Fishery tax*	*Commercial tax*	*Fishery tax*
Sakitsu 崎津	27	465	0	465
Imatomi 今富	0	0	0	0
Ooe 大江	6	40	24	40
Takahama 高浜	45	0	177	50
Kozatoko 小田床	33	0	42	0
Simotsufukae 下津深江	18	0	33	0
Tororo 都呂々	6	0	0	0
Fukuregi 福連木	0	0	0	0
Total amount	135	505	276	555

Sources: Ueda House Archive 4–42; "Sho-unjo-mono narabini Nou-kan-syo-kasegi sono-hoka kaki-age-cho (= Tax Registers) 諸運上物并農間諸稼其外書上帳," Onoue Monjo 尾上文書, reprinted in Mataji Miyamoto 宮本又次, "Tenryo Amakusa no Shogyo to Tonya 天領天草の商業と問屋," in *Kyushu Bunka-shi Kenkyu-sho Kiyou* 九州文化史研究所紀要, 2 (1951), 75.

Table 2.2 Increase in the forms of transportation in Takahama

Year	*Population*	*Horses*	*Cattle*	*Boats and ships*
1714	1,272	109	16	13
1732	1,804	134	8	7
1750	2,400	135		10
1761	2,557	138		10
1772	2,875	125		18
1789	3,078	117		12
1821	3,559			
1838	3,638	578		58
1861	3,732	632	16	81

Source: Ueda House Archive, 4–4, –5, –7, –10, –12, –14, –19, 7–78, 12–22, 4-additional-1–2.

continue to be locally mined today. Ceramic stones from Takahama were also exported to ceramics production areas in Kyushu famous for Arita-Yaki (有田焼, Arita ceramics). Horses were presumably used for transportation of ceramic stones and other materials from mountain areas to the seaport.

From the basis of the rice subsistence economy, not only did the agricultural development of polyculture in Japan during this period produce a complex history of commercialization and ecological management, new "cropping decisions raised difficult problems of work scheduling, which were in some degree different for each farm family," and required the anticipation of new work flows for an entire growing season, because "each crop entailed a number of narrowly timed tasks: seed treatment, soil preparation, planting (and often transplanting), repeated and numbered weeding and fertilizations, and so on."[47]

These busy, full-time farming practices combined with those of other industries represent a new type of "industriousness," different from the kind required for concentrated rice production, and also different from the market-oriented "industriousness" described by de Vries: The European "industriousness" occurred first "in those peasants households that could follow the course of specialization by concentrating household labor in marketed food production." Second it was "revealed most clearly by proto-industrial and proletarian households" forced to work "longer and harder." Finally, it was observed in "rural non-agricultural households, seasonal migrants and even urban residents" as "seasonal peaks, and the intensification of agriculture to supply marketed foodstuffs, particularly bread grains, intensified the demand for harvest labor."[48]

Saito insists that the "industrious revolution" in the sense in which it is referred to by Jan de Vries never occurred in early modern Japan. There certainly existed a tendency for peasant families to work longer, but they were never willing to substitute the purchase of commodities at markets for the production of those goods at home.

> Needless to say, the demand by the farm family for cash incomes increased over time, so did their actual cash earnings and perhaps their cash spending

> as well. However, it is not unlikely that a substantial proportion of increased cash earnings were used to buy cash fertilizers and other input materials in order to raise yields.[49]

Otherwise the livestock was especially important for natural fertilizer, because feces and urine of horses and cattle were mixed with cut grasses trampled by them in the cattle shed or horse barn. Making natural fertilizer was the main use of livestock, which required grasslands in the village surroundings mostly managed by the village administration collectively.[50]

No conversion from "horse" (livestock) power to "man" power took place in its entireness in Japan. "A recent survey of evidence on this issue has revealed that Hayami's interpretation of the Nobi evidence is questionable, and that the man-cattle ratio varied considerably from region to region, as well as from time period to time period."[51] While in Owari (尾張) the livestock per capita, household, and village decreased drastically between 1670 and 1820, in some villages such as Takahama and in Tamana (玉名), Higo (肥後), an enormous amount of horsepower came to be used in a later period of Tokugawa Japan. Other evidence derived from my own research reveals that: (1) the number of horses and cattle also increased in Tado-county (多度郡), Marugame (丸亀) in Shikoku (四国), in the southwestern part of Japan, where there was a population of not only horses but also cattle; (2) a similar mixed population was observed in Aki-county (安芸郡), Tosa (土佐) in Shikoku; but in some villages in other parts of Japan there were only horses, especially in villages of northern and cooler central mountainous part of Japan,[52] whereas in other villages, such as Harima (播磨), located in the southwestern part of Japan, only cattle were found (Table 2.3).

Though we can find an enormous diversity of livestock arrangements, it is almost certain that many more working hours were demanded of humans, with or without the use of livestock, when monoculture agriculture was retained, or when polycultural farming had to be extended and combined with commercialized industries for the sustainability of the villages.

Neither cotton nor meat markets were developed in early modern Japan. A market economy fostered by "industriousness" did exist, but it was much weaker than that in northwestern Europe. Saito summarized commercialization in Japan compared to that in Europe:

> In short, while in north-western Europe both "Smithian"[53] and capitalist forces were at work (with the former being mainly in manufacturing and the latter in agriculture and commerce), Tokugawa growth was nothing but a "Smithian" process. As a consequence, in the European case, the wealth created through capitalist developments contributed to expanding the "extent of the market" in the domestic economy, which in turn acted as a factor fostering the further division of labor, whereas the whole process of Japan's pre-modern growth was somewhat more restricted.[54]

Beef, pork, chicken, and lamb markets did not develop in early modern Japan, and capital investments for livestock occurred only on a small scale for the special production of horses in limited regions such as Morioka (盛岡). Not only did capital investment for livestock fail in Japan (see Table 2.3), but historical research in various fields regarding the nutrition basket (see Table 2.4) also shows a definitive difference between Japan and the West.

The nutrition basket was created for the purpose of comparing living standards internationally. It reveals a clear cultural difference in nutrition patterns, which developed variously along with the art and volume of production. The basic patterns were essentially formed in Tokugawa Japan.

But to talk about "industriousness" only in the sphere of markets, especially the human labor market,[55] limits the discussion. "Industriousness" in early modern Japan applied not only to commercialized work by family businesses, but also to voluntary and disciplinary work by social management, which was strongly combined with the utilization of livestock power.

Conclusion

The competing notions of "industriousness"—the impetus to freely participate in the market versus the obligation to work hard in given callings—seem to have few common elements for comparison. The former notion of "industriousness" fostered commercialization and prepared the ground for the Industrial Revolution. The latter, in contrast, limited the expansion of individual market participation within the possibility of labor forces in family business.

The pursuit of a family business was the social norm in Japan. It was not the effort of individuals but the professionalized capability of the family that was responsible for the lives of family members and also for the wealth of the village. This was not only the social norm, but also the prevalent approach to business management, which could establish and change the constellation of kinship and non-kinship relations. Members of a family were members of the taskforce for the business. The division of family properties between individuals was unthinkable. Such family businesses protected and disciplined the individuals within the family, and thus proletarization was an exceptional phenomenon. The family was a management body; therefore, members of a family were expected to work hard within the family's assigned professional calling.

From a comparative view of historical sources, the crucial issue is that historical sources such as "probate inventories" in Europe never appeared in Tokugawa Japan. Instead, annual population registers such as the "*Shumon Aratame Cho*," which registered the household or family as a unit certificated to be non-Christians, were archived, in many cases for more than 100 years in the house of the village headman, *shoya*. The number of livestock was also registered in other surveys which were combined with those religious registers. Such registrations were a necessary product for managing human capital and livestock as a collective labor force within a village. T. C. Smith characterized the household and family priority precisely, asserting that "time was not a personal possession but belonged primarily to families and, through them, to kin, neighbors, and villages."[56] His argument should be extended to livestock arrangements.

Table 2.3 Livestock per capita, per household, and per village in Tokugawa Japan

	Livestock per capita	*Livestock per household*	*Livestock per village*	*Note*
Owari 尾張, 635 villages, 1670–1820	0.05 (1670)–0.01 (1820)	0.28 (1670)–0.06 (1820)	19.43 (1670)–6.61 (1820)	Horses and cattle
Tamana 玉名, Higo 肥後, 8 villages, 1633–1880	–	–	24.38 (1633)–51.25 (1880)	Horses and cattle
Takahama 高浜, 1714–1868	0.04 (1789)–0.21 (1838)	0.36 (1789)–1.19 (1838)	117.00 (1789)–758.00 (1838)	Horses, mainly
Tado-county 多度郡, Marugame 丸亀, 9 villages, 1698—around 1845	0.07 (1698)–0.12 (ar. 1845)	0.41 (1698)–0.49 (ar. 1845)	78.14 (1698)–144.86 (ar. 1845)	Horses and cattle
Saku 佐久, Shinano 信濃, 6 villages, 1798	–	0.40	–	Horses and no cattle
Harima 播磨, 6 villages, 1737–1749	–	0.13	–	Cattle and no horses
Aki-county 安芸郡, Tosa 土佐, 108 villages, 1743	0.09	0.41	38.39	Horses and cattle

Sources: Hayami (2003), pp. 311–312, 314 for Owari; Saito (2004), p. 157 for Tamana; original data for Takahama, Ueda House Archive, 4–4, –5, –7, –10, –12, –14, –19, 7–78, 12–22, 4-additional-1–2; Kihara (2003), pp. 141–142 for Tado-county; Mizumoto (2003), p. 67 for Saku and Harima; original counting from Hirao (1952), pp. 35–42 for Aki-county.

Table 2.4 European and Asian baskets and nutrition (per person/year)

	Japan		*India*	*Europe*
	Type A: Fish/rice	*Type B: Cereals*		
Foods				
Bread (kg)	–	–	–	208.0
Peas (l)	4.0	4.0	52.0	52.0
Meat (kg)	–	–	26.0	26.0
Butter (India = ghi) (kg)	–	–	10.4	10.4
Soybean (kg)	52.0	26.0	–	–
Rice (kg)	114.0	30.0	143.0	–
Barley or wheat (kg)	10.0	70.0	–	–
Fish (kg)	3.5	–	–	–
Cereal (kg)	16.0	75.0	–	–
Cooking oil (l)	1.0	1.0		
Non-foods				
Oil (l)	5.0	5.0	5.0	5.0
Cloth (m)	2.6	2.6	2.6	2.6
Daily calories: about 1940 kilo calorie				
Daily protein: 80 g				

Source: Allen (2005), p. 115, Table 5.1 "European and Asian baskets and nutrition," and pp. 125–128; Bassiono and Ma (2005); Saito (2008), p. 107.

The most decisive point concerns the diversity of livestock arrangements in Tokugawa Japan (see Table 2.3). We cannot find any stereotype of the combination of livestock selected in each village. Livestock was used for cultivation, making natural fertilizer, and conveyance, but never for meat production. Natural climatic and agricultural conditions and commercialization patterns produced various forms of relationships between humans and animals. Hayami's "Industrious Revolution" represents only one pattern of such diversity, because there were various possibilities in the "industriousness" of each village of Tokugawa Japan.

Changing attitudes towards nature were diversified from village to village, and from family to family according to callings and the natural conditions of each village. But a much more crucial point is that the interactive learning process addressing "nature" began among the almost entire Japanese farmer population in accordance with local and regional market expansion from the peaceful seventeenth century. The market decided the future and continuity of the family, including the possession of livestock as part of the labor force and as a compost maker.

Environmental diversity engendered the differentiated household and livestock system, because the family business tried to pursue the best performance of the family and its labor force constellation. The variety and differentiated formations of livestock composition had their origins in agricultural and commercial performance of each village based on ecological and market conditions.

Thus, Hayami's original arguments should be extended to encompass "environmental history," i.e., the history of interactions between humans and nature.

This discussion opens new questions: Was the "industrious revolution" in the West, in the sense intended by de Vries, also accompanied by a kind of changing attitude towards nature? Or, did significant ecological changes in the West occur only after the "Industrial Revolution"? And how can we reconcile our understanding of the interactive history of humans and nature in the West and in other parts of the world with the debates about the "industrious revolution"?

Acknowledgment

This study was a revised version of a paper which was presented as a keynote speech titled "Two types of 'industriousness' and changing attitudes towards nature in early modern times. An international comparison" at the Second Conference of East Asian Environmental History held at National Dong Hwa University, College of Management (Management Building), Hualien, Taiwan, R.O.C., on October 24–26, 2013. It was supported by the Japan Society for the Promotion of Science Grants-in-Aid for Scientific Research (A) (19203018) from April 2007 to March 2011 and Grants-in-Aid for Scientific Research (B) (21401002). I am also grateful to the Eurasia Project, which was conducted by Professor Akira Hayami from April 1995 to March 2000 and funded by the Japanese Ministry of Education, Culture, Sports, Science, and Technology, and I was also supported by the fellowship of the Rachel Carson Center of the University of Munich from July to September 2013 to finish this paper.

Notes

1 Murayama Satoshi 村山聡, "Work Hard as a Family Business in a Peaceful and Calm Village in Early Modern Japan," *GEO-COMMUNICATION* Working Paper No. 8. March 11, 2014, 12–13. See Keith Thomas, *Man and the Natural World: Changing Attitudes in England 1500–1800* (London: Penguin Books, 1984) for the environmental history of poor horses in England; we also have a recent contribution by Kanehira Kenji 兼平賢治, "A Life Course of Early Modern Horses Viewed from Horse (南部馬にみる近世馬の一生)," in Mizumoto Kunihiko 水本邦彦 (ed.), *Human Businesses and Early Modern World* [人々の営みと近世の自然] (Tokyo: Yoshikawa Kobun-kan, 2013), pp. 134–163. These environmental history arguments combine with the claim of Takei Koichi 武井弘一 in "Development of New Rice Fields and Early Modern Ecological System (新田開発と近世型生態系)," in Mizumoto Kunihiko (ed.), *Human Businesses and Early Modern World*, p. 206, for the emergence of an early modern ecological system would be discussed in future.

2 Sheilagh Ogilvie, "Consumption, Social Capital, and the 'Industrious Revolution' in Early Modern Germany," *Journal of Economic History* 70.2 (2010): 289; Frederick Duke of Würtemberg, "A True and Faithful Narrative of the Bathing Excursion," in William Brenchley Rye, *England as Seen by Foreigners in the Days of Elizabeth and James I* (London: John Russell Smith, 1865), pp. 7, 14.

3 Robert C. Allen, *The British Industrial Revolution in Global Perspective* (Cambridge: Cambridge University Press, 2009), pp. 12–13; Sir James Steuart, *Inquiry into the Principles of Political Economy* (1767), pp. 53–54, 58, 199, 229.

4 T. C. Smith, "Peasant Time and Factory Time in Japan," in T. C. Smith, *Native Sources of Japanese Industrialization, 1750–1920* (Berkeley, Los Angeles, London: University of California Press, 1988), pp. 205–206.
5 T. C. Smith, "Peasant Time and Factory Time in Japan," p. 199: Smith translated a Japanese original metaphoric sentence: "Koin Oshimubeshi 光陰惜しむべし" (lightness = sun and darkness = moon should be valued), into an English rational one: "it must treat time as precious."
6 Ibid.
7 Hayami Akira 速水融, "The Emergence of the Economic Society and its Characteristics (経済社会の成立とその特質)," in Socio-Economic History Society 社会経済史学会 (ed.), *In Search of a New Image of the Edo Period* [新しい江戸時代像を求めて] (Tokyo: Toyo Keizai Shinposha, 1967), pp. 3–18.
8 Saito Osamu 斎藤修, "An Empirical Re-examination of the Industrious Revolution (勤勉革命論の実証的再検討)," *Mita Gakkai Zasshi* [三田学会雑誌] 97 (April 2004): 151–152; Hayami Akira, "The Emergence of the Economic Society and its Characteristics"; Akira Hayami, "A Great Transformation: Social and Economic Change in Sixteenth and Seventeenth Century Japan," *Bonner Zeitschrift für Japanologie*, No. 8 (1986): 3–13; Hayami Akira, "The Industrial Revolution vs. The Industrious Revolution (産業革命対勤勉革命)," in Hayami Akira (ed.), *Economic Society of Early Modern Japan* [近世日本の経済社会] (Tokyo: Reitaku University Press, 2003), pp. 307–322.
9 Jan De Vries, "The Industrious Revolutions in East and West," in Gareth Austin and Kaoru Sugihara (eds.), *Labour-Intensive Industrialization in Global History* (London and New York: Routledge, 2013), p. 72.
10 Akira Hayami, "A Great Transformation: Social and Economic Change in Sixteenth and Seventeenth Century Japan," p. 6.
11 Hayami Akira, *Historical-demographic Studies of Early Modern Villages* [近世農村の歴史人口学的研究—信州諏訪地方の宗門改帳分析], (Tokyo: Toyo Keizai Shinposha, 1973); Hayami Akira, *Land Survey and Peasants in Early Modern Times* [近世初期の検地と農民] (Tokyo: Chisen Shoin, 2009); Tomobe Kenichi 友部謙一, *Peasant Household Economies in Pre-industrial Japan: Subjective Equilibrium and Market Economy* [前工業化期日本の農家経済—主体均衡と市場経済] (Tokyo: Yuhi-kaku, 2007), pp. 4–6; Furushima Toshio 古島敏雄, *Agricultural Structure of Early Modern Japan* [近世日本農業の構造] (Tokyo: Tokyo Daigaku Syuppan-kai, 1957).
12 Tomobe Kenichi, *Peasant Household Economies in Pre-industrial Japan*, pp. 42–46.
13 Jan De Vries, "The Industrial Revolution and the Industrious Revolution," *Journal of Economic History* 54 (1994): 249–269.
14 Craig Muldrew, *Food, Energy and the Creation of Industriousness: Work and Material Culture in Agrarian England, 1550–1780* (Cambridge: Cambridge University Press, 2011), p. 14.
15 Jan De Vries, *The Industrious Revolution: Consumer Behavior and the Household Economy, 1650 to the Present* (Cambridge: Cambridge University Press, 2008), p. 273.
16 Kaoru Sugihara, "The East Asian Path of Economic Development: A Long-term Perspective," in Giovanni Arrighi, Takeshi Hamashita, and Amark Selden (eds.), *The Resurgence of East Asia: 500, 150, and 50 Year Perspectives* (London: Routledge, 2003), pp. 78–103; Kaoru Sugihara, "The State and the Industrious Revolution in Tokugawa Japan," *Working Papers in Economic History*, No. 02/04 (London School of Economics, 2004); Gareth Austin and Kaoru Sugihara (eds.), *Labour-Intensive Industrialization in Global History* (London and New York: Routledge, 2013).
17 Jan De Vries, "The Industrial Revolution and the Industrious Revolution," p. 249; Jan De Vries, *The Industrious Revolution.*
18 Hayami Akira, "The Industrial Revolution vs. The Industrious Revolution," pp. 312–313. See also Osamu Saito, "Wages, Inequality, and Pre-Industrial Growth in

Japan, 1724–1894," in Robert C. Allen, Tommy Bengtsson, and Martin Dribe (eds.), *Living Standards in the Past: New Perspectives on Well-Being in Asia and Europe* (Oxford and New York: Oxford University Press, 2005), pp. 77–97.

19 Osamu Saito, "Pre-Modern Economic Growth Revisited: Japan and the West," in London School of Economics, *Working Papers in Economic History*, No. 16/05 (June 2005): 1–2; T. C. Smith, *Native Sources of Japanese Industrialization.*

20 Jan De Vries, "The Industrious Revolutions in East and West," in Gareth Austin and Kaoru Sugihara (eds.), *Labour-Intensive Industrialization*, p. 77.

21 Osamu Saito, "Pre-Modern Economic Growth Revisited," p. 31.

22 Alan Macfarlane, "The Invention of the Modern World," Spring-Summer Serial 2012, Chapter 1 of 17, available online: http://fortnightlyreview.co.uk/2012/04/invention1/, accessed on August 10, 2013.

23 Ogilvie, "Consumption, Social Capital, and the 'Industrious Revolution'."

24 See Osamu Saito, "Pre-Modern Economic Growth Revisited," and "Wages, Inequality, and Pre-Industrial Growth in Japan, 1724–1894," p. 86. Nakamura Satoru 中村哲, *Underlying Structure of the Meiji Restoration: The Origin of Japanese Capitalism Formation* [明治維新の基礎構造ー日本資本主義形成の起点] (Tokyo: Miraisya, 1968).

25 The *shoya* in Takahama kept diaries dealing with the governance of a village community. The Ueda (上田) house archive contains 88 diaries. Of these, we were able to investigate those kept by Yoshiuzu (宜珍), who served as *shoya* for 30 years (1789–1818). Our research project has digitized these diaries, making available a total of 1,849 pages and more than one million characters in a book format. This digitization has allowed us to use the diaries analytically, because we can access the information easily and systematically. These diaries were not intended to preserve memories, but functioned as documentation, manuals for administration, and evidence for the government in order to sustain a village community safely and peacefully.

26 Kitou Hiroshi 鬼頭宏, *Japanese History Viewed from the Population* [人口から読む日本の歴史] (Tokyo: Kodansha, 2000).

27 Moriyuki Oe, "Problems and Implications of Japan's Aging Society for Future Urban Developments," *Policy and Governance Working Paper Series*, No. 89 (March 2006).

28 Tomobe Kenichi, *Peasant Household Economies in Pre-industrial Japan*, pp. 37–42.

29 Mizumoto Kunihiko 水本邦彦, "Early Modern Humans and Nature (人と自然の近世)," in Mizumoto Kunihiko (ed.), *Human Businesses and Early Modern World*, pp. 8–12.

30 Aiso Kazuhiro 相蘇一弘, "Water Supply of Modern Hineno Who Saw from Pictorial Map (絵図から見た近世日根野の水利)," *Japanese History Research* [日本史研究] 310 (1988): 57–79.

31 Takei Koichi, "Development of New Rice Fields and Early Modern Ecological System," p. 195; Kigoshi Ryuzo 木越隆三, *Studies of Land Survey of Nobunaga and Hideyoshi and Kokudaka* [織豊期検地と石高の研究] (Kyoto: Katsura-shobo, 2000).

32 Mizmumo Kunihiko, "Early Modern Humans and Nature," p. 13.

33 T. C. Smith, *Native Sources of Japanese Industrialization, 1750–1920*, p. 137.

34 Mizmumo Kunihiko, "Early Modern Humans and Nature," p. 14.

35 Ibid., pp. 14–15.

36 T. C. Smith, *Native Sources of Japanese Industrialization, 1750–1920*, p. 173.

37 Ibid., p. 176.

38 Ibid., p. 181.

39 Ibid., p. 181.

40 Ibid., p. 185.

41 Ibid. (Nagatsune), p. 185.

42 Mizumoto Kunihiko, *Early Modern Times from the View of Grass Mountains* [草山の語る近世] (Tokyo: Yamakawa Syuppan-sha, 2003).

43 Saito Osamu 斎藤修, "Economic History of Population and Environment (人口と環境の経済史)," in Sawa Takamitsu 佐和隆光 et al. (eds.), *Economic Theory of Environments* [環境の経済理論] (Tokyo: Iwanami Shoten, 2002), p. 239.

44 Higashi Noboru 東昇, "Products Survey by Domains: Souvenirs, Local Specialties, and Presents (藩の産物調査と土産・名物・献上)," in Mizumoto Kunihiko (ed.), *Human Businesses and Early Modern World*, pp. 223–245.

45 See Note 25.

46 In comparison to Takahama, rice production in a village such as Niremata (楡俣), which was populated by 672 people in 1812 (Bunka 2), had a village *kokudaka* of 1,570 *koku*, representing tax payments that had not changed since 1623 (Genna 元和 9). This amounted to 2.34 *koku* per capita. Niremata was located in an extraordinarily fertile but disaster-prone area with flooding risks as a result of being in the downstream area of Nobi's (濃尾) three big rivers, the Kiso (木曽), Nagara (長良), and Ibi (揖斐). See Narimatsu Saeko 成松佐恵子, *Social Conditions and Life Viewed from Diaries of Shoya* [庄屋日記にみる江戸の世相と暮らし] (Kyoto: Minerva Shobo, 2000), pp. 1–10.

47 T. C. Smith, *Native Sources of Japanese Industrialization*, pp. 206–209.

48 Jan De Vries, "The Industrious Revolutions in East and West," pp. 79–80.

49 Osamu Saito, "Pre-Modern Economic Growth Revisited," pp. 38–39.

50 Isoda Michifumi 磯田道史, "Agricultural Development of the 17th Century: Using Grasses and Cattle (17世紀の農業発展をめぐって—草と牛の利用から—)," *Japanese History Research* [日本史研究] 402 (1996): 27–50; Isoda Michifumi, "Agriculture and Agricultural Politics of the Domain in the Period of Early Modern Village Formation (近世村落成立期の農業と藩農政)," in Research Association of Okayama-Domain 岡山藩研究会 (ed.), *Consciousness and Relations of Domain World* [藩世界の意識と関係] (Tokyo: Iwata-shoin, 2000), pp. 43–72.

51 Osamu Saito, "Pre-Modern Economic Growth Revisited: Japan and the West," p. 35.

52 Kanehira Kenji, "A Life Course of Early Modern Horses Viewed from Horse."

53 For Smithian growth, see J. Mokyr, *The Lever of Riches: Technological Creativity and Economic Progress* (New York: Oxford University Press, 1990), pp. 4–6, where he identified four processes of economic growth. The other three are Solovian, Boserupian, and Schumpeterian. Osamu Saito, "Pre-Modern Economic Growth Revisited," pp. 3–4.

54 Osamu Saito, "Pre-Modern Economic Growth Revisited," p. 43.

55 Osamu Saito, "Wages, Inequality, and Pre-Industrial Growth in Japan."

56 T. C. Smith, *Native Sources of Japanese Industrialization*, p. 201.

Bibliography

Aiso Kazuhiro 相蘇一弘. "Ezu kara mita Kinsei Hineno no Suiri 絵図からみた近世日根野の水利 (Water Supply of Modern Hineno Who Saw from Pictorial Map)," *Nihonshi Kenkyu* 日本史研究 [*Japanese History Research*], 310 (1988).

Allen, Robert C. "Real Wages in Europe and Asia: A First Look at the Long-Term-Patterns," in Robert C. Allen, Tommy Bengtsson, and Martin Dribe (eds.), *Living Standards in the Past: New Perspectives on Well-Being in Asia and Europe*. Oxford and New York: Oxford University Press, 2005, pp. 111–130.

Allen, Robert C. *The British Industrial Revolution in Global Perspective*. Cambridge: Cambridge University Press, 2009.

Allen, Robert C., Tommy Bengtsson, and Martin Dribe (eds.). *Living Standards in the Past: New Perspectives on Well-Being in Asia and Europe*. Oxford and New York: Oxford University Press, 2005.

Austin, Gareth and Sugihara Kaoru (eds.). *Labour-Intensive Industrialization in Global History*. London: Routledge, 2013.

Bassiono, J.-P. and D. Ma. "Japanese Unskilled Wages in International Perspective, 1741–1913: A Comparison with Europe," *Research in Economic History*, 23 (2005): 229–248.

De Vries, Jan. "The Industrial Revolution and the Industrious Revolution," *Journal of Economic History*, 54 (1994): 249–269.

De Vries, Jan. *The Industrious Revolution: Consumer Behavior and the Household Economy, 1650 to the Present*. Cambridge et al.: Cambridge University Press, 2008.

De Vries, Jan. "The Industrious Revolutions in East and West," in Gareth Austin and Sugihara Kaoru (eds.), *Labour-Intensive Industrialization in Global History*. London and New York: Routledge, 2013, pp. 65–84.

Furushima Toshio 古島敏雄. Kinsei Nihon Nogyo no Kozo 近世日本農業の構造 [*Agricultural Structure of Early Modern Japan*]. Tokyo: Tokyo Daigaku Syuppan-kai, 1957.

Hayami Akira 速水融. "Keizai Shakai no Seiritsu to sono Tokushitsu 經濟社會の成立とその特質 (The Emergence of the Economic Society and its Characteristics)," in Socio-Economic History Society (Shakai Keizaishi Gakkai 社会经济史学会) (ed.), *Atarashii Edo Jidaizo o Motomete* 新しい江戸時代史像を求めて [*In Search of a New Image of the Edo Period*]. Tokyo: Toyo Keizai Shinposha, 1967, pp. 3–18.

Hayami Akira 速水融. *Kinsei No-son no Rekishi Jinkou-gaku teki Kenky—Shin-shu Suwa-chiho no Syumon-Aratame-Cho Bunseki* 近世農村の歴史人口学的研究—信州諏訪地方の宗門改帳分析 [*Historical Demography of Early Modern Villages: An Analysis of Syumon-Aratame-Cho in Shinsyu-Suwa Region*]. Tokyo: Toyo Keizai Shinpo-sha, 1973.

Hayami Akira. "A Great Transformation: Social and Economic Change in Sixteenth and Seventeenth Century Japan," *Bonner Zeitschrift für Japanologie*, No. 8 (1986): 3–13.

Hayami Akira 速水融. "Sangyo Kakumei Tai Kinben Kakumei 産業革命対勤勉革命 (The Industrial Revolution vs. The Industrious Revolution)," in Akira Hayami, *Kinsei Nihon no Keizai Shakai* 近世日本の经济社会 [*Economic Society of Early Modern Japan*]. Tokyo: Reitaku University Press, 2003, pp. 307–322.

Hayami Akira 速水融. *Kinsei Shoki no Kenchi to Noumin* 近世初期の検地と農民 [*Land Survey and Peasants in Early Modern Times*]. Tokyo: Chisen Syoin, 2009.

Higashi Noboru 東昇. "Han no Sanbutsu Chosa to Miyage—Meibutsu—Kenjyo (Products Survey by Domains: Souvenirs, Local Specialties, and Presents)," in Kunihiko Mizumoto 水本邦彦 (ed.), *Hitobito no Itonami to Kinseino Shizen* 人々の営みと近世の自然 [*Human Works and Early Modern Nature*]. Tokyo: Yoshikawa Kobun-kan, 2013, pp. 223–245.

Hirao Michio 平尾道雄. "Tosa-han Gou-son Chosa-sho 土佐藩郷村調査書 (Village Inquiries of Tosa-Domain)," *Tosa Shidan*土佐史談 [*Tosa Local History Journal*], 81 (1952): 34–42.

Isoda Michifumi 磯田道史. "17-seiki no Nougyo-hatten wo megutte: Kusa to Ushi no Riyo kara 17世紀の農業發展をめぐって一草と牛の利用から (Agricultural Development of the 17th Century: Using Grasses and Cattle)," *Nihon-shi Kenkyu* 日本史研究 [*Japanese History Research*], 402 (1996): 27–50.

Isoda Michifumi 磯田道史. "Kinsei Sonraku Seiritsu-ki no Nougyo to Han-Nousei 近世村落成立期の農業と藩農政 (Agriculture and Agricultural Politics of the Domain: Mountains and Fields)," in Okayama-han Kenkyu-kai 岡山藩研究会 (Research Association of Okayama-Domain) (ed.), *Consciousness and Relations of Domain World*. Tokyo: Iwata-shoin, 2000, pp. 43–72.

Kanahira Kenji 兼平賢治. "Nanbu-uma ni miru Kinsei-ba no Issho 南部馬にみる近世馬の一生 (A Life Course of Early Modern Horses Viewed from the Nanbu Horse)," in

Kunihiko Mizmumo (ed.), *Hitobito no Itonamito Kinsei no Shizen* 人々の営みと近世の自然 [*Human Works and Early Modern Nature*]. Tokyo: Yoshikawa Kobun-kan, 2013, pp. 134–163.

Kigoshi Ryuzo 木越隆三. *Syokuho-kenchi to kokudaka no kenkyu* 織豊期検地と石高の研究 [*Studies of Land Survey of Nobunaga and Hideyoshi and Kokudaka*]. Kyoto: Katsura-shobo, 2000.

Kihara Hiroyuki 木原溥幸. *Chiiki ni miru sanuki no kinsei* 地域にみる讃岐の近世 [*Early Modern Sanuki Viewed from Regions*]. Takamatsu: Biko-sha, 2003.

Kitou Hiroshi 鬼頭宏. *Jinko kara yomu Nihon no Rekishi* 人口から読む日本の歴史 [*Japanese History Viewed from the Population*]. Tokyo: Kodansha, 2000.

Macfarlane, Alan. "The Invention of the Modern World," Spring-Summer Serial 2012. Available online: http://fortnightlyreview.co.uk/2012/04/invention1/, accessed on August 10, 2013, Chapter 1 of 17.

Mizumoto Kunihiko 水本邦彦. *Kusa-yama no Kataru Kinsei* 草山の語る近世 [*Early Modern Times from the View of Grass Mountains*] Tokyo: Yamakawa Syuppan-sha, 2003.

Mizumoto Kunihiko 水本邦彦. "Hito to Shizen no Kinsei 人と自然の近世 (Early Modern Humanities and Nature)," in Kunihiko Mizmumo (ed.), *Hitobito no Itonamito Kinsei no Shizen*人々の営みと近世の自然 [*Human Works and Early Modern Nature*]. Tokyo: Yoshikawa Kobun-kan, 2013, pp. 8–38.

Mokyr, J. *The Lever of Riches: Technological Creativity and Economic Progress*. New York: Oxford University Press, 1990.

Muldrew, Craig. *Food, Energy and the Creation of Industriousness: Work and Material Culture in Agrarian England, 1550–1780*. Cambridge: Cambridge University Press, 2011.

Murayama Satoshi 村山聡. "Work Hard as a Family Business in a Peaceful and Calm Village in Early Modern Japan." *GEO-COMMUNICATION* Working Paper No. 8. March 11, 2014. Online: rfweb.ed.kagawa-u.ac.jp/project/wiki/.../wiki.cgi?...

Nakamura Satoru 中村哲. *Meiji ishin no kiso kozo. Nihon Shihon-syugi Keisei no Kiten* 明治維新の基礎構造—日本資本主義形成の起点 [*Underlying Structure of the Meiji Restoration: The Origin of Japanese Capitalism Formation*]. Tokyo: Mirai-sha, 1968.

Narimatsu Saeko 成松佐恵子, *Shoya Nikki ni Miru Edo no Seso to Kurashi* 庄屋日記にみる江戸の世相と暮らし [*Social Conditions and Life Viewed from Diaries of Shoya*]. Kyoto: Minerva Shobo, 2000.

Oe, Moriyuki. "Problems and Implications of Japan's Aging Society for Future Urban Developments," *Policy and Governance Working Paper Series*, No. 89, March 2006.

Ogilvie, Sheilagh. "Consumption, Social Capital, and the 'Industrious Revolution' in Early Modern Germany," *Journal of Economic History*, 70:2 (2010): 287–325.

Saito Osamu 齋藤修. "Jinko to Kankyo no Keizai-shi 人口と環境の經濟史 (Economic History of Population and Environment)," in Sawa Takamitsu 佐和隆光 et al. (eds.), *Kankyo no Keizai Riron* 環境の經濟理論 [*Economic Theory of Environments*]. Tokyo: Iwanami Shoten, 2002, pp. 209–240.

Saito Osamu. "Kinben kakumei-ron no jisshoteki sai-kennto 勤勉革命論の実証的再検討 (An Empirical Re-examination of the Industrious Revolution)," *Mita Gakkai Zasshi*, 97 (April 2004): 151–161.

Saito, Osamu. "Pre-Modern Economic Growth Revisited: Japan and the West," Working Papers in Economic History, No. 16/05. London School of Economics, June 2005.

Saito, Osamu. "Wages, Inequality, and Pre-Industrial Growth in Japan, 1724–1894," in Robert C. Allen, Tommy Bengtsson, and Martin Dribe (eds.), *Living Standards in the*

Past: New Perspectives on Well-Being in Asia and Europe. Oxford and New York: Oxford University Press, 2005, pp. 77–97.

Saito Osamu. *Hikaku Keizai Hatten-ron. Rekishi-teki Approach* 比較経済発展論–歴史的アプローチ– [*Comparative Studies of Economic Development: A Historical Approach*]. Tokyo: Iwanami-shoten, 2008.

Sugihara Kaoru. "The East Asian Path of Economic Development: A Long-term Perspective," in Giovanni Arrighi, Takeshi Hamashita, and Amark Selden (eds.), *The Resurgence of East Asia: 500, 150, and 50 Year Perspectives*. London: Routledge, 2003, pp. 78–103.

Sugihara Kaoru. "The State and the Industrious Revolution in Tokugawa Japan." Working Papers in Economic History, No. 02/04, London School of Economics, 2004.

Takei Koichi 武井弘一. "Shinden Kaihatsu to Kinsei-gata Seitai-kei 新田開発と近世型生態系 (Development of New Rice Fields and Early Modern Ecological System)," in Kunihiko Mizmumo (ed.), *Hitobito no Itonamito Kinsei no Shizen* 人々の営みと近世の自然 [*Human Works and Early Modern Nature*]. Tokyo: Yoshikawa Kobun-kan, 2013, pp. 194–222.

Tomobe Kenichi 友部謙一. *Zen Kougyouka-ki Nihon no Noka Keizai. Shutai Kinko to Sijyo Keizai* 前工業化期日本の農家経済—主体均衡と市場経済 [*Peasant Household Economies in Pre-industrial Japan: Subjective Equilibrium and Market Economy*]. Tokyo: Yuhi-kaku, 2007.

Smith, Thomas C. "Farm Family By-employments in Preindustrial Japan," in T. C. Smith, *Native Sources of Japanese Industrialization, 1750–1920*. Berkeley, Los Angeles, London: University of California Press, 1988, pp. 71–102.

Smith, Thomas C. "Okura Nagatsune and the Technologists," in T. C. Smith, *Native Sources of Japanese Industrialization, 1750–1920*. Berkeley, Los Angeles, London: University of California Press, 1988, pp. 173–198.

Smith, Thomas C. "Peasant Time and Factory Time in Japan," in T. C. Smith, *Native Sources of Japanese Industrialization, 1750–1920*. Berkeley, Los Angeles, London: University of California Press, 1988, pp. 199–235.

Thomas, Keith. *Man and the Natural World: Changing Attitudes in England 1500–1800*. London: Penguin Books, 1984.

Part II

Case studies of mainland China and Taiwan

3 Water management and the legitimization of the Yongle reign, 1403–1424

An approach of political ecology

Cho-ying Li

Introduction

From the first to the third year of the Yongle reign (1403–1405), Xia Yuanji (夏原吉, 1366–1430) was in charge of water management in the Lower Yangtze Delta. This chapter explores how his appointment and the hydraulic measures he employed were related to the Yongle Emperor's (永樂皇帝, r. 1403–1424) anxious desire to legitimize his reign after he had overthrown the Jianwen Emperor (建文皇帝, r. 1399–1402). The numerous measures the Yongle Emperor applied to legitimize his new reign after he had successfully "quelled the disturbances (*jingnan* 靖難)" have often been discussed in academia.[1] For instance, Benjamin A. Elman has analyzed how the Yongle Emperor created a national canon to liberate him from the image of a usurper, and to lend him the reputation of a sage-king instead.[2] Hok-Lam Chan has examined how the Yongle Emperor rewrote the records about the Hongwu (洪武, r. 1368–1398) and the Jianwen reigns to make it look as if he was the legitimate heir to the throne.[3] Zhang Xiaoping and Zhou Shaoqing have demonstrated how this strong monarch with weak legitimacy employed a wide variety of methods to change this original predicament.[4]

The research outlined above has offered us a thorough knowledge of the Yongle Emperor's efforts towards, and impact on, institutions, historical discourses or Classical orthodoxy. But some other measures the Yongle Emperor undertook immediately after his ascension to the throne, and the way they were related to his political legitimacy, have been mostly neglected in academia. The example I wish to explore in this chapter is Xia Yuanji's hydraulic work. In discussing this matter, scholars have mostly emphasized the Yongle Emperor's efforts to safeguard his people's livelihood.[5] However, the question of how these efforts were linked to legitimizing his reign awaits examination. As this chapter will show, Xia Yuanji's appointment, his choice of hydraulic measures, the discussions about sage-kingship he touched upon in the process, as well as the Yongle Emperor's attitude towards water management and the way it changed, were all shaped by considerations about political legitimacy. It is therefore worthwhile to explore the topic from this perspective.

In addition, this chapter will discuss another implication. In recent years, research on water management in the Lower Yangtze Delta has illustrated that the setup of hydraulic management there changed significantly between the Song (960–1278) and the Ming (1368–1643) Dynasties. While there had originally been a network of major channels, there was now a smaller, but more complex, network of minor waterways. The original spacious polders were abandoned in favor of installations that worked with small polders. And instead of draining all the water into the Wusong River (吳淞江), it was now guided out of the region through the channels on the north and south sides of this river.[6]

All this is very convincing. But if we focus on Xia Yuanji's discussions on water management, as well as on his concrete actions, we cannot but notice that he emphasized one tenet: namely that the local waterways had to be managed in accordance with the theory of the "Three Rivers (三江)." Even though the *Tribute of Yu* (*Yugong* 禹貢) reads, "After the Three Rivers were channelled into the sea, the Marsh of Zhen [today's Lake Tai 太湖] was stabilized (三江既入, 震澤底定)," Wang Jian-ge points out that the strategies water managers employed in the Song and Yuan (1280–1367) Dynasties still focused on dredging the Wusong River itself, as well as the lakes and major channels on its periphery. They channeled the water of these lakes and waterways into the Wusong River, and afterwards led the water from the Wusong River into the sea. So the theory that the Three Rivers channeled the water into the sea only existed in the imagination of the water managers. In reality there were absolutely no efforts to dig channels towards the southeast and northeast, and to try and recover the other two rivers.[7] In other words, Xia Yuanji's arguments and actions contradicted the usual hydraulic practices that had been in use since the Song and the Yuan. So what was Xia Yuanji's motivation?

To answer this question, this chapter will draw upon the historical background, that is, the Yongle Emperor's urgent desire to legitimize his reign immediately after his ascension to the throne, to offer a tentative explanation. What was happening was an interplay between the practical demands of politics (Yongle's expectations and Xia Yuanji's pressure to accomplish something great in the new reign), the cultural significance with which the Classics had loaded the "Three Rivers," and the environmental conditions at the time (these resulted from a combination of the natural environment and efforts undertaken by earlier generations). All this shaped the historical uniqueness of the decisions that Xia Yuanji made.

This chapter is greatly indebted to discussions going on in political ecology. Paul Robbins' *Political Ecology: A Critical Introduction* is widely praised as the most comprehensive book in the field, in terms of historical depth and theoretical reflections.[8] In his discussion of the "conceptual and methodological challenges" political ecology faces, Robbins dedicates a section to explaining why political ecologists argue that what we call "nature" is a man-made construction.[9] He gives the example of the forest: A forest is not a natural rather than a social "phenomenon, object or idea." It is "forged by convention and context," and, since we have become used to it, "enforced by its very taken-for-grantedness." Furthermore, when the decision whether or not a "bunch of trees" is regarded as a "forest," touches upon the question whether "significant state and international resources will be invested in its protection or its eradication,"[10] the matter becomes highly political. Therefore, Robbins points out, political ecologists are particularly concerned with the following targets: "things that are most taken for granted"; "concepts and objects that, by being constructed one way, disallow alternative interpretations, and so mask political motivations and activities"; and "those constructions that contribute to constructing the objects of the world."[11] In sum, political ecologists want to uncover the process due to which natural objects are regarded and treated as "natural," and how they produce political or social effects in the human world.

Such an approach is relevant to the analysis of Xia Yuanji's hydraulic management in many ways. First of all, in reality the Lower Yangtze Delta was not in urgent need of water management at the time. There were several local incidents, but not really significant enough to prompt the state to action. However, the Yongle Emperor magnified them to the degree that the entire region had to engage in water management. And by responding quickly and extensively to these local requests, he implemented his political motive, which was to legitimize his reign, without saying so explicitly.

Second, the discourse of the "Three Rivers" that Xia Yuanji had worked out was put into practice not because it best reflected the network of waterways in the Lower Yangtze Delta, but because its cultural implications perfectly suited the political expectations of the time, and because it was more convenient for Xia, who wanted to achieve remarkable accomplishments quickly, to make use of already existing environmental conditions. Under such circumstances the "Three Rivers" became the obvious blueprint for water management, and the natural object for hydraulic work.

Third, in a time of such political expectations, polder dykes and "pedal pump rescue (車救)" also emerged as two methods of flood control and as the mechanism to organize local officials and people in hydraulic management. This was designed to glorify the Yongle reign as a government of grand capability that could reshape the institutions. And, finally, after similar political hopes for legitimacy had prompted the Yongle Emperor to consider the possible negative effects such measures could have on politics, he abandoned water management, even though it had originally looked like the only reasonable course of action. Instead, he started dealing with flood disasters by granting tax exemption and famine relief. This demonstrates clearly that the water management in the early Yongle years was not a necessity induced by nature, but rather a man-made strategy, which appeared and disappeared depending on how its political and social effects were assessed.

Xia Yuanji in the first year of the Yongle reign

On 17 July 1402, the Yongle Emperor usurped the throne.[12] Over one month later, on 23 August 1402, Xia Yuanji was promoted from Right Vice Minister of Revenues to Left Vice Minister.[13] Then, on 30 September 1402, he was again promoted, this time to Minister of Revenues.[14] After another seven months, he was appointed to take charge of hydraulic management in the Lower Yangtze

Delta.[15] The political significance of his speedy promotion and appointment should be examined against the background that the Yongle Emperor was eager to stabilize his regime by threatening and conciliating the officials of the previous reign, shortly after he had overthrown the Jianwen Emperor.

Let us first look at his threats. After the Prince of Yan had conquered Nanjing, he made a public announcement on 13 July 1402, in which he condemned twenty-seven people as "false courtiers and treacherous ministers (左班奸臣)." He proclaimed: "Those who are guilty I do not dare to pardon. Those who are innocent I do not dare to execute (有罪者, 予不敢赦; 無罪者, 予不敢殺)." Later on, he made another list of twenty-three people, whom he called "treacherous and evil ministers (奸惡官員)."[16] Before Xia Yuanji's promotion to Right Vice Minister of Revenues, thirty of the people, whose names had been on these two lists, had already been executed or they had committed suicide. Among them were Huang Zicheng (黃子澄, ?–1402), Fang Xiaoru (方孝孺, 1357–1402)[17] and Lian Zining (練子寧, ?–1402).[18]

With respect to conciliation, the Yongle Emperor had once expressed his intention to treat the old officials with leniency. For instance, after Chen Ying (陳瑛, ?–1411) had been recalled as Right Vice Censor-in-chief in the summer of 1402, he suggested that six officials of the previous reign should be disgraced posthumously. The Yongle Emperor, however, answered that he had even "forgiven and employed (宥而用之)" the people who had been listed as treacherous and evil ministers, let alone those who had never been ranked among them. The Yongle Emperor said: "When they were in his [i.e., the Jianwen Emperor's] service, they dedicated themselves to him with all their hearts. You must not question this! (彼食其 [指建文皇帝]祿, 自盡其心耳, 悉勿問!)"[19] Moreover, later on in the same year, he commanded Xie Jin (解縉, 1369–1415) to carefully read more than a thousand secret memorials submitted by Jianwen's ministers and to keep those concerning military and fiscal matters, but burn the whole of the rest that was offensive. In other words, the Yongle Emperor had no intention of making any thorough inquiries into the matter.

It was not just two different sorts of people that were reconciled or threatened by the Yongle Emperor. In fact, one and the same person could experience both. After the Emperor had ordered the burning of over a thousand secret memorials of the Jianwen Reign, he also calmly inquired of people like [Xie] Jin: "You all must have submitted some." This statement must have made high officials who had served during both reigns nervous. So as a reaction, they kowtowed and were unable to reply.[20] But then the Yongle Emperor stated that "Yesterday you all served him, so you were loyal to him. Today you serve me, so you must be loyal to me. (爾等前日事彼則忠於彼, 今日事朕當忠於朕)." That is, during the Jianwen Reign, one should pledge loyalty to the Jianwen Emperor, and manage the affairs of the state as if one was in the Emperor's place, but now a new reign had been established, the officials of this reign must be absolutely loyal to the Yongle Emperor, and decide over the great affairs of the country in his best interest. This explains that threats and conciliation were two sides of the same coin. And both sides could be simultaneously used on one and the same man.

Under these circumstances, Xia Yuanji was not in as dangerous a position as all those officials whose names had been written on the two lists. But his situation was still very serious. First of all, when the Jianwen Emperor was being overthrown, he was just then inspecting Qizhou (蕲州, i.e., Qichun 蕲春 in today's Hubei Province) as an investigation commissioner. And later someone "arrested him to present him to the Prince of Yan (執公以獻燕王)."[21] At that time, many people were in similar situations. And many of them were killed because they refused to serve Yongle. For example, the investigating censor Gan Lin (甘霖, ?–1402) was arrested, pleaded for his death and was executed.[22]

Xia's submission allowed him to escape death for now. But it also made him an imperial courtier and subjected him to the rules of the new reign. His surrender alone would not guarantee his safety in the future. This becomes clear from the example of Sheng Yong (盛庸, ?–1403). Sheng had offered strong resistance against the Prince of Yan's campaign to the south, and caused him extreme trouble. After the capitulation, he was first commanded to guard Huai'an (淮安). And shortly afterwards, he was transferred to Shandong. But from the beginning to the end, Sheng Yong always felt unsafe. In the end, Chen Ying impeached Sheng in a memorial and as a result in 1403 he "was expropriated of his noble ranks, sent to prison and died, all his sons were killed as well (削爵下獄死, 諸子皆被戮)."[23]

Therefore, Xia Yuanji obviously could not relax, just because he was left vice minister, or minister, of Revenues. We know that he had been serving as an official since the Hongwu years. Because both Hongwu and Jianwen Emperors had thought highly of him, and that high-ranking senior officials had put him into important positions, he had gained a very good reputation over the past ten years in the Ministry of Revenues.[24] This was certainly one reason why the Yongle Emperor promoted him. But it must have also put him under pressure to make a contribution to the new reign. This indicates that we should take a closer look at some of the elements in Xia Yuanji's performance in the first years of the Yongle era.

According to the accounts in the *Veritable Records of Taizong* [太宗實錄], Xia Yuanji's first task was to close down the (old-style) equitable exchange of rice for salt in the (whole) realm, and to concentrate on the equitable exchange in Beiping (i.e., today's Beijing). This was designed to solve the problem of grain shortage for all garrisons in Beiping. But all the procedures were decided by the Yongle Emperor; he did not have the chance to display his excellent abilities.[25] But later on, the Yongle Emperor would approve two of the concrete measures Xia Yuanji had suggested in the context of the equitable exchange of rice for salt.[26]

However, in the same month in which he had been promoted to minister, Xia Yuanji's first advice, which concerned the resettlement of refugees, was rejected by the Yongle Emperor. At that time, the Emperor had commanded to have all refugees of war transfer back to their hometowns. But after listening to a lot of people's complaints, Xia Yuanji thought their misery deserved pity, so he suggested:

> Let them stay in the capital for now, and to have them wait until we investigate the situation of their hometowns. If they still have relatives there, we can send them home. If they do not [have any], we should follow their requests. In both cases, justice would be done to their human feelings. (請命暫居京師, 俟行勘原籍, 有親即遣歸, 果無, 則宜從所言, 亦人情兩便).

But the Yongle Emperor did not find their statements credible. He said:

> Which human feeling does not yearn for home? Those who have taken them captive and who have bought them must have put these words [of complaint] into their mouths. They are not credible. [The refugees] can stay in the capital for the moment, and we will give them grain rations. And we will wait, and send them back home when it is warmer in the spring. If they have been bought as brides with betrothal gifts, we have to make investigations too. If it is true, we grant them [their request]. (人情孰不懷土? 是必虜買之家教誘為此言, 不可盡信. 姑留在京, 給與口糧, 待春暖遣歸, 如果用財禮聘娶, 亦須行勘. 有實, 然後給之).[27]

What appears from this reply is the Yongle Emperor's view that Xia Yuanji had been deceived by the people. Therefore, the Yongle Emperor only accepted part of his suggestion, but did not make the slightest compromise about the general direction of the whole policy.

Two months later, Xia Yuanji put forward a suggestion about the woodblocks used for printing paper money:

> The blocks for printing paper money, used by the supervisorate of paper money, are very old. The seal characters are fading. Besides, they all read "Hongwu reign." Next year, we are starting the Yongle reign. So we should change them. (寶鈔提舉司鈔板, 歲久, 篆文銷乏, 且皆洪武年號, 明年改元永樂, 宜併更之).

But the Yongle Emperor disagreed and said:

> If the woodblocks need replacement because they are old, we replace them. [But] we do not need to change "Hongwu" into "Yongle." I obey all the laws that Taizu has established. Using "Hongwu" forever would be fine as well. (板, 歲久, 當易則易, 不必改洪武為永樂. 蓋朕所遵用皆太祖成憲, 雖永用洪武可也).[28]

This shows that Xia Yuanji did not yet understood the Yongle Emperor's plan to justify his usurpation by honoring Taizu's institutions.[29]

If we sum up Xia Yuanji's performance in these six months, we can say that the Yongle Emperor approved of the way Xie managed concrete affairs. But Xia Yuanji had apparently not fully understood the intentions behind some of the Yongle Emperor's policies. When the Yongle Emperor appointed Xia Yuanji to water management in the Lower Yangtze Delta, he was, on the one hand, hoping that the even more senior Yu Xin (郁新, 1346–1405) would take over a wide diversity of duties of the Ministry of Revenues. But on the other hand, he was confident in Xia Yuanji's abilities to concentrate all his efforts on a specific task—flood control in the Lower Yangtze Delta. Furthermore, when the Yongle Emperor instructed him that he saw no need to exchange the blocks used for printing paper money, Xia Yuanji must have understood the Yongle Emperor's urgent wish to legitimize his reign. (See the next section for more detail.) Therefore, for Xia Yuanji, the water management in the Lower Yangtze Delta presented not only an opportunity to perform well in the new reign, but also a duty to undertake practical actions that would help Yongle to gain political legitimacy.

Hydraulic management as a strategy and practice to establish political legitimacy

This section seeks to demonstrate that Xia Yuanji's hydraulic management involved multiple rearrangements and that all these rearrangements were designed to consolidate the political legitimacy of the Yongle reign, by showing that it was a capable government concerned with its people's welfare. I will explore three facets of these rearrangements in this section.

1 Renarrating: The Yongle Emperor seized the opportunity presented by the fact that local people were renarrating hydraulic management, and responded to their problems quickly, actively and broadly. In this way, he illustrated that he was better than the Jianwen Emperor and that he was more capable of increasing the popular welfare.
2 Restructuring: When Xia Yuanji chose the "Three Rivers" as the matrix for his speedy and extensive water management, he made best use of the existing environmental conditions, and gave the Yongle Emperor the image of a sage-king. For this purpose, he restructured the hydraulic network of that region on a large scale.
3 Reorganizing: Xia Yuanji did not only dredge out the waterways. He also established separate mechanisms for preventing floods and for draining off flood water: polder dykes and "pedal pump rescue." In these two systems, the choice of responsible officials and their capacity to mobilize the people was particularly emphasized. And as the government reorganized the way people were mobilized, it was able to use their labor efficiently, and to benefit the people on the basis of the strengthened hydraulic regulations.

Renarrating hydraulic management

Was it very urgent for the Yongle Emperor to send Xia Yuanji to the Lower Yangtze Delta to conduct hydraulic management there? According to the local gazetteers, the Prefectures of Jiaxing (嘉興), Suzhou (蘇州) and Songjiang

(松江) did not suffer from inundations during the Hongwu and Jianwen reigns.[30] Even if we consult books on water management, which contain rather complete records of floods, we can see as well that hardly any inundations happened during the long thirty years of the Hongwu and Jianwen reigns.[31] Particularly noteworthy is that after the Hongwu Emperor had commanded the students of the Imperial Academy to go to the prefectures and counties of the realm, and to supervise officials and people in improving the hydraulic condition in 1394, no inundations happened for eight years.[32] In other words, the flood control of the Jianwen reign had been very successful. So why did the Yongle Emperor regard water management in the Lower Yangtze Delta as urgent, when the hydraulic situation there was perfectly fine?

The documents mentioned above convey the impression that hydraulic management had been quite satisfactory in the Lower Yangtze Delta before the Yongle Emperor succeeded to the throne. Still, in late 1402, a native of Jiading (嘉定) County, Zhou Cheng (周程, ?–?), submitted a petition, in which he claimed that the Lower Yangtze region was suffering over and over because of the siltation of the Wusong River. Therefore he asked the emperor to send a high minister to "survey the land, to plan the works, and to have the people dredge the Wusong River in the slack season of farming (丈量計工, 令於農隙之時, 役民疏鑿)."[33]

There are several important points in this petition. First, it requests action over a very large region known as "Eastern Wu (東吳)" which is roughly the area south of the Yangtze River in today's Jiangsu. Second, it implies that natural disasters were regular phenomena. Third, it asks to send an official to deal with the situation; that is, it insinuates that measures undertaken previously had not been effective. According to the two chapters, "Hydraulic Management (*Shuili* 水利)" and "Abnormal Events (*Xiangyi* 祥異)," in the *Gazetteer of Jiading County* (嘉定縣志), floods and droughts were by no means regular phenomena in the early Ming.[34] Nevertheless, Zhou Cheng's petition shows that, in the eyes of ordinary people, the siltation of the Wusong River had long become the source of endless worries. In other words, the hydraulic situation in the Lower Yangtze Delta in the first thirty odd years of the Ming is reported differently in the various collections of documents on hydraulic management and in the local gazetteers on the one hand, and in Zhou Cheng's petition on the other. This reflects how much official evaluations and actions differed from the people's experience and opinion.

The collections of documents on hydraulic management and the local gazetteers drew upon official documents and data. So they only recorded those disasters which the officials thought were worth recording or that deserved intervention.[35] Consequently the records give the impression that nothing happened for many years, which means in fact that no floods occurred that were big enough to result in official action. But the people who lived in the drainage area of the Wusong River did not share the official view. In Zhou Cheng's eyes, the inundations were already big in scope and frequency of occurrence. The descriptions the officials gave in their records and the people in their petitions can all be regarded as facts from their respective points of view. The important question is, the facts from which viewpoint did the Yongle Emperor, who had only been on the throne for a few months, let determine his actions.

Three months after Zhou Cheng's petition had arrived, the Yongle Emperor sent Xia Yuanji to the Lower Yangtze Delta to take charge of hydraulic management. This shows his decision to accept Zhou Cheng's view of the facts, and to act accordingly. It was a highly political action that the Yongle Emperor did not adopt the definitions that came from the bureaucracy, but made his decisions on the basis of popular opinion instead. The reason is that Xia Yuanji's appointment got across at least two political messages, which were related to establishing legitimacy. First, although, by official standards, the situation described by Zhou Cheng was insufficient to result in official intervention, the Yongle Emperor responded immediately to the request. This displayed him as a monarch who could quickly rescue the people from their hardship. Second, Zhou Cheng claimed that the people had been suffering from floods for a long time. For the Yongle Emperor, this implied a criticism of the mediocrity and incompetence of the monarch and the officials of the Jianwen reign. And when he had a high official embarking on extensive hydraulic management, he showed, by contrast, that he, this new emperor, was better at enhancing the popular welfare.

In fact, the central request in Zhou Cheng's petition also appears in the *Veritable Records of Taizong*, in an outline that describes the whole story of Xia Yuanji's appointment:

> On the *jiyou* day, [the emperor] commanded the minister of Revenues, Xia Yuanji, to manage the water of all the prefectures of Zhexi. At the time, Jiaxing, Suzhou and Songjiang Prefectures had been suffering from floods in consecutive years. Many edicts had commanded the officials to deal with the situation, but without any effect. Therefore, this order was issued. (己酉, 命戶部尚書夏原吉往浙西諸郡治水. 時, 嘉興蘇松諸郡頻歲水患, 屢勑有司督治, 訖無成績, 故有是命).[36]

To be sure, the *Veritable Records of Taizong* were compiled in the Xuanzong reign (1398–1435). And the remarks after the words "at the time" were also written by the compiler. But a comparison of Zhou Cheng's petition and these remarks here shows the compiler's description—Jiaxing, Suzhou and Songjiang Prefectures had been suffering from floods in consecutive years. Many edicts had commanded the officials to deal with the situation, but without any effect. This means that Zhou Cheng's portrayal of the present situation and of his previous experience provided a context for the Yongle Emperor, in which he could engage in actions which would show that he was different from the Jianwen Emperor, and, most of all, that he was better than the Jianwen Emperor and had more reason to succeed to the throne. Most significantly, when he responded to this memorial by appointing Xia Yuanji, he did not limit the scope of his actions to Suzhou. Instead, he extended his intervention to three prefectures. This proactive and extensive handling of people's requests shows how anxious the Yongle Emperor was to create the impression that his reign was one that could tackle problems comprehensively and that was capable and competent. Under these circumstances, Xia Yuanji's water management in the Lower Yangtze region was also burdened with the Yongle Emperor's expectation to legitimize himself after his usurpation of the throne.

How fervently the Yongle Emperor hoped to achieve this goal is mirrored in the fact that he kept sending officials to assistant Xia Yuanji. He had only appointed Xia Yuanji in the fourth month. But when he got news of a "prolonged rainfall)" in the Lower Yangtze Delta in the sixth month, he said to the Left Vice Minister of Revenues, Gu Pu (古朴, ?–1428): "In Suzhou, Songjiang and Jiaxing, the lakes must be overflowing. We should quickly send someone to see what is going on. (蘇松嘉湖水必泛濫, 宜速遣人視之)." Hence the Right Vice Minister of Revenues, Li Wenyu (李文郁, fl. 1390s–1420s), was sent as an assistant to Xia Yuanji.[37] After two months, the Yongle Emperor was again "worried about floods in Suzhou and Songjiang (以蘇松水患為憂)." So he ordered the Right Assistant Censor in Chief, Yu Shiji (俞士吉, ?–1435) to take the *Collections on Water Management* [水利集] to the Lower Yangtze Delta and let Xia Yuanji consult it, in the hope that Xia would "explore a method to solve and manage [the situation] (講究拯治之法)."[38] In the following year, he first made the Vice Minister of the Court of Judicial Review, Yuan Fu (袁復, ?–1406), as Xia Yuanji's assistant.[39] And later, he sent Song Xing (宋性, ?–1424), the Administration Vice Commissioner of Shaanxi, to take over from Yuan Fu.[40] At the same time, Xia Yuanji was also supported by the Left Vice Commissioner of the Bureau of Transmissions, Zhao Juren (趙居任, ?–1419), and by supervising secretary of the Office of Scrutiny for Works, Yao Boshan (姚伯善, *juren* 舉人, degree in 1399).[41] Furthermore, the Yongle Emperor was looking for ways to select talent from among the people. For instance, Ye Zongxing (葉宗行, ?–?) from Huating (華亭) and Zhao Youtong (趙友同, 1364–1418) from Jinhua (金華) were appointed to join in Xia Yuanji's works.[42] By sending so many assistants to Xia Yuanji, the Yongle Emperor may certainly have had the intention to pool their wisdom and help him resolve the problems. But it also shows his anxiety and the pressure Xia Yuanji must have been under.

Another example that shows the Yongle Emperor's desire to legitimize himself through hydraulic management is his treatment of Tang Zong (湯宗, ?–1427), the Prefect of Suzhou. On 20 August 1403, that is, four months after Xia Yuanji's appointment, Tang Zong was "sent to jail, because he had sat back and watched the flood (以坐視水患下獄)."[43] If we look at everything that Tang Zong had done, this seems to be an unfounded accusation. When he discovered that, due to the floods, "the common people had become homeless and had fallen into arrears by over a million *dan* [of rice], and that they had no way of paying it (細民流離, 逋租百餘萬石無所出)", he did not cruelly force them to make the payments. Instead he turned to the rich people and ordered them to pay the rice on their behalf. The local rich people "knew that he could care for people, and they all followed his order. And before the third month, all the debts had been paid. (知其能愛民, 皆從其令. 不三月逋負悉完)."[44] In fact, Tang Zong did not "sit back and watch the flood." So how can we explain his accusation? One possible explanation would be that Tang Zong's measures did not earn the Yongle Emperor the people's gratitude. The reason why the rich people agreed to hand over their rice was that they acknowledged Tang Zong as an official who cared for the people. And when the poor people were relieved from the burden of rent payment, they were also grateful to Tang Zong. In other words, Tang Zong showed benevolence to the lower reaches of society, but he did not give the credit to the top. This thwarted the Yongle Emperor's plans to gain political legitimacy by dealing with the floods. Moreover, when the Yongle Emperor sent Xia Yuanji to this region, he was hoping to get the floods under control through hydraulic measures, rather than by bringing aid relief after the disaster had happened. Tang Zong put his energy into finding a solution for the default of payments. Even though this problem had been caused by the floods, he did not take any steps at hydraulic control. In the eyes of the Yongle Emperor, this was indeed "sitting back and watching the floods." So it was only natural that he should be punished.

Restructuring the river system

Tang Zong's arrest must have been a warning for Xia Yuanji. Not only did he have to focus his policies on dealing with the flood, but also on achieving a result as quickly as possible. He also had to be intent on ascribing his achievements to the Yongle Emperor. On the day on which Tang Zong went to prison, Xia Yuanji presented his plan. After it had been approved, he first performed a sacrificial ceremony for the deity of the river before he put the plan into practice. What deserves a detailed analysis is how his notions about hydraulic management and the creation of the Yongle Emperor's image as a sage-king were revealed in his plan, as well as in the benediction he declaimed at the sacrifice for the river spirit. Let us first look at the "Benediction for the Sacrifice to the River Deity (祀告河神文)":

> Thinking retrospectively of the Three Rivers, they were dug by the Holy Yu. They have been left neglected for an extensive period of time, and are now about to dry out. Thus paddy fields are damaged and the spirit of the River Deity is withering. Fortunately, now the sage has re-emerged in the world. Day and night he is worried about, and dedicating himself to, [the people's welfare]. He ordered me to drain and guide [the water]. I plead with you, the River Deity, for your help to make the water go back into its proper path. (溯維三江, 神禹開鑿, 曠久不治, 涓流將涸, 農田被害, 神氣蕭索; 睠茲九重, 聖人復作, 宵旰憂勤, 命予疏瀹, 願神助襄, 水歸其壑).[45]

This text starts out with telling how a sage-king, the Great Yu, dug the channels for the Three Rivers. It goes on to recount that the Three Rivers had not been managed for a long time, and that the people's property has been damaged because water conservancy had been neglected. Then the tone changes to define the Yongle Emperor's ascension to the throne as the "re-emergence of a sage-king." And it describes how hard the Yongle Emperor was working for the affairs of the country, and how he has commanded him, Xia Yuanji, to come here and undertake water management. This description compares the Yongle Emperor to the Great Yu, the sage-king of antiquity, who blessed the common people with his benevolence. It also announces that the emperor was going to thoroughly change this state of affairs, in which the government was idle and inactive. This echoes perfectly the distorted description of the Jianwen reign, which we find in Xia Yuanji's order, and it implies that he was going to bring a new spirit and a new welfare to the people. Moreover, Xia Yuanji made it clear that he was coming under the "order" of the Yongle Emperor. So all his achievements would be ascribed to the Yongle Emperor's incessant "worries and dedication, day and night."[46]

Furthermore, the Yongle Emperor's commitments and considerations did not only have an effect on the human world, but could even reach out into the realm of the spirits. Xia Yuanji constructed the two expressions "the spirit of the River Deity is withering" and "I plead with you, the River Deity, for your help" as a dichotomy between present and past. In this way he emphasized that, if a monarch did not take care of the human affairs, the River Deity would be ruined as well. Now that the Yongle Emperor was determined to make up for the failures of the previous reign, the River Deity should also, like an echo, assist in "making the water go back into its proper path." In sum, the central message of the *Benediction* read: The Yongle Emperor was a sage-king, who could appoint a high official, summon the spirits, and who would bless the people with a peaceful and prosperous life.

The "Three Rivers" were an important medium through which Xia could convey this message. According to the *Benediction*, they were not natural waterways, but channels that had only come into being because they had been dug by human labor. They ran smoothly and plentifully, not due to nature, but because they were maintained by human efforts. In other words, the hydrography constructed through the "Three Rivers" gave the monarch and his official a starting point for their intervention. Moreover, whether they were in a condition that was harmful or useful to the people became a criterion by which one could distinguish a weak, incompetent and mediocre monarch, from a sage-king who cared about the sorrows of his people. In view of the aforementioned political ecology, we should regard the "Three Rivers" not as a "natural" thing, but as a "political" object, which emerged in response to specific expectations and interests. When Xia Yuanji submitted his plan to the Yongle Emperor, he therefore used the "Three Rivers" as the framework for his hydraulic project. He said:

> I have consulted with my fellow officials and experts on hydraulic management, and we have formulated an outline [for this project].... The solution lies in dredging the Wusong River and all the channels [around it], to discharge the silt, and to guide [the water] into the sea.

We know that the tidal waves make the Wusong River ... clog up with sand and mud. Whenever we dredge it out, it silts up again, and this is not the way to obtain a lasting effect. It is about a hundred and twenty *li* from the Long Bridge in Wujiang County to the Xiajie Channel. Although it is said that [the Wusong] is dredged clear [along this route], it is [in fact] shallow and narrow at many spots. From the Xiajie Channel to a point south to Pukou of Shanghai County, it is certainly over one hundred and thirty *li*. The tidal sand is silting it up, water chestnuts and reeds are growing thickly. And so [this section of the Wusong] has already become dry land. If we wanted to dredge it out, the expenses for the works would be massive. Moreover, the sand and the mud that are being washed up are floating around everywhere. So it would be difficult to conduct the works.

My assistants and I had investigated the situation, and we learnt that the Liujia Channel, that is, the ancient Lou River, in Jiading [County] flows directly into the sea; and that the Baimao Channel in Changshu empties directly into the Yangtze. These are all big waterways, and their flows are rapid. So we should dredge the channels on the northern and southern shores of the Wusong, like the Anting. Then we can manipulate all the water of Lake Tai into the Liujia and the Baimao Channels. [And from there] we can guide them directly into the Yangtze or the sea. Moreover, the Huangpu [River] of Songjiang is the main waterway that leads into the Wusong [River]. But these days, its lower reaches have silted up and it flows slowly. So [we should] dredge it out. Next to it, there is the Fanjia Creek, which flows up to Nanqiang River Mouth, and leads directly into the sea. We should dredge it out, so that it gets deep and broad, link it up to the Huangpu, and then [use it to] drain out the water of Lake Mao. All this is the old track through which the Three Rivers flow into the sea, as it is described in the *Tribute of Yu*.

臣與共事官屬及諳曉水利者參考與論得其梗概 ... 拯治之法, 要在濬滌吳松諸浦港, 泄其壅遏, 以入于海). 按, 吳松江 ... 當潮汐之衝, 沙泥淤積, 屢濬屢塞, 不能經久. 自吳江長橋至下界浦約百二十餘里, 雖云疏通, 多有淺窄之處; 自下界浦抵上海縣, 南蹌浦口, 可百三十餘里, 潮沙壅障, 菱蘆叢生, 已成平陸. 欲即開濬, 工費浩大, 且灩沙游泥泛泛動盪, 難以施工. (臣等相視得嘉定之劉家港即古婁江, 通大海, 常熟之白茆港, 徑入大江. 皆係大川, 水流濬急. 宜濬吳松南北兩岸安亭等浦港, 以引太湖諸水入劉家、白茆二港, 使直注江海. 又松江之大黃浦乃通吳松要道, 今下流壅塞難流疏濬. 傍有范家濱至南蹌浦口, 可徑達海, 宜濬令深闊, 上接大黃浦以達泖湖之水。此即《禹貢》三江入海之跡.[47]

At the beginning of this text, Xia Yuanji starts with a rough description of his water management by first referring to the Wusong River. But, drawing upon an on-site examination, he explains immediately that he would encounter two difficulties if he targeted all his efforts on the Wusong River—the expenses for the works would be massive and it would be difficult to conduct the works. Then he mentions a few waterways, which, he thinks, should be dredged instead. And in the end, he concludes that "all this is the old track through which the Three Rivers flow into the sea, as it is described in the *Tribute of Yu*." In other words, Xia Yuanji's first choice would have been the Wusong River. But practical difficulties forced him to find an alternative plan. However, it was apparently not enough to back up his argument with a mere description of the difficulties that would arise out of the expenses and the execution of the work. Consequently, he emphasized that these alternative waterways were exactly what was called the "Three Rivers." This gave Xia Yuanji's blueprint for water control a foundation in the Classics, because the *Tribute of Yu*, that is, the passage "after the Three Rivers were channelled into the sea, the Marsh of Zhen was stabilized," expresses the central idea of how the Great Yu tamed the floods.

However, there is a great ambiguity in the Classics and their commentaries as to what the Three Rivers actually means. Even though many Southern Song texts specialized on the *Tribute of Yu* were written, they do not offer any substantial help in defining to what exactly the "Three Rivers" refer.[48] This gave Xia Yuanji flexibility for his interpretation. And it also gave him space to formulate his reading in a way that would benefit his project. In the Hongwu reign, people such as the Hanlin (翰林) scholar Liu Sanwu (劉三吾, 1313–1399?) wrote the *Collected Selection of the Commentaries of the Book of Documents* [書傳會選], with the aim to draft a standard book that could be used for the civil service examinations. As their primary source, they used the *Collected Commentaries of the Book of Documents* [書經集傳] by the Song scholar Cai Shen (蔡沈, 1167–1230). And they observed the following principle when putting the *Collected Selection of the Commentaries of the Book of Documents* together:

> We keep all [the information] that is in accordance with Cai's commentaries, but we do not denounce any others in a biased fashion. We change whatever is contradictory to Cai's commentaries, but we do not defend him as if we were holding a sectarian view. (凡蔡傳之合者存之, 不預立意見以曲肆詆排. 其不合者則改之, 亦不堅持門戶以巧為回護).[49]

Consequently, they adopted Cai Shen's opinion that

> [A]fter seventy *li*, the [Wu]song River splits up. The [waterway] that flows into the sea in the north-east is called the Lou River. And the one that flows into the south-east is called the Dong River. Together with the [Wu]song River, they are called the Three Rivers. (松江下七十里分流東北入海者為婁江, 東南流者為東江, 併松江為三江).[50]

In the early Ming Dynasty, however, these efforts to standardize the Classics and their commentaries did not have any particular power to restrain Xia Yuanji's flexibility. Although Cai Shen's teacher, Zhu Xi (朱熹, 1130–1200), did not have a definite opinion on what the "Three Rivers" were, he had stated once that he tended to agree with Xue Jixuan (薛季宣, 1134–1173), and did not trust the definition of Su Shi (蘇軾, 1037–1101). Zhu Xi explained why he held this view:

> [Su] Dongpo has never personally seen the rivers in the south-east region. He only says this based on his imagination.... Xue Shilong [i.e., Xue Jixuan] says that the Three Rivers empty into the sea after flowing from the Marsh of Zhen. I suspect that he [Xue] has seen the rivers in the south-east region and thus I feel that he is right. (東坡不曾親見東南水勢, 只是意想硬說,... 士龍說震澤下有三江入海. 疑他曾見東南水勢, 說得恐是).[51]

With these words, Zhu Xi asserted that an explanation would be more valuable if one had seen the locality with one's own eyes and conducted an on-the-spot examination, than if one reached a conclusion by merely speculating on various commentaries or by making things up oneself. So Xia Yuanji's on-site investigation of the hydrographic system in the Lower Yangtze Delta gave him enough authority to cast off the restraints imposed upon him by the *Collected Selection of the Commentaries of the Book of Documents*, and enabled him to follow the reference he had found for the Three Rivers himself.

Although Xia Yuanji had a considerable degree of freedom when interpreting the "Three Rivers," he did not choose the Liujia Channel (also called Liujiahe 劉 (or 瀏) 家河), the Baimao Channel (also called Baimaotang 白茆塘) and the Huangpu River (in addition to the Fanjia Creek 范家濱) arbitrarily. On the contrary, this decision was shaped by the pressure he had to accomplish something quickly, and it was limited by (and, at the same time, it also benefitted from) the hydraulic discourses and measures of the previous dynasties, and by the effect these discourses and measures had had on the environment.[52]

First of all, Xia Yuanji was not the first to notice that the Wusong River was difficult to manage and that one had to find alternative waterways; nor was he the first to discuss this point in writing. In the Yuan Dynasty, Zhou Wen-ying (周文英, 1265–1334) once mentioned in the mid-1320s that people had piled up stones to construct a dyke by the Long Bridge on the Wusong River. Although this improved traffic between two sides of the river, it made the drainage paths under the stone dyke too small to drain off the water of Lake Tai. Moreover, the dyke made the river flow slowly and softly, so that it could not wash away the tidal silt. As a result, the siltation of the lower course of the river became a serious problem. He vehemently advocated abandoning the idea of dredging the Wusong River. His first choice for a waterway was the Liujia Channel. So he made the following recommendation:

> South of the Liujia Channel is a big waterway called the Nanshiqiao Channel. It has been naturally broad and deep in recent years, and it flows directly into the Liujia Channel. At this time, there are two battalion commissioners, from the Yang and Fan boat families, who supervise seagoing boats of three to five thousand *liao* [unit of lumber] that are riding at anchor there.... In its middle reaches, it [the Nanshiqiao Channel] flows past a ridge, and then runs south-west to connect the Heng Channel, the Marshes of Guo and the Zhang Creek. Finally, it flows into the Wusong River, together with the Xiajia Channel and the Wengzi Channel. The distance

> between the two points—[each is on one side of the ridge]—where the water is deep is about thirty or fifty *li* [1 *li*=one-third of a mile]. Between these two points are many interconnected smaller channels, which at the moment are shallow and, at some places, winding and narrow. If we dredged them [i.e., the shallow, winding and narrow channels], they will become deep and broad, and allow seagoing ships of several hundred *liao* to sail directly to Fengmen [a gate of Suzhou City]. This would be a great route along which we could drain off the water of Lake Tai. (劉家港南有一大港名曰南石橋港, 近年天然闊深, 直通劉家港. 見有船戶楊千戶、范千戶等三、五千料海船於此灣泊 … 於中正過堈身, 西南通橫塘、郭澤、張涇, 以至夏駕浦、奣子港入吳淞江. 水深處相去約三、五十里, 中間通連小港, 即目水淺, 間有迂迴窄狹, 若使疏濬深闊, 可行數百料海船, 直抵葑門, 則太湖泄水一大路也.)[53]

This shows that Zhou Wen-ying was hoping to connect several waterways, in order to guide the water of Lake Tai from the upper reaches of the Wusong River into the Liujia Channel, and ultimately to manipulate them into the Yangtze. Although he acknowledged that there were a few spots between the Liujia Channel and the Wusong River that were winding and narrow, he thought it was worth investing some manpower into these places and trying to improve them. Once the two waterways were connected, they would be deep and broad enough for ships of several hundred *liao* to sail on them. This would also be a significant contribution to flood control in the Lake Tai basin.

It was no accident that Zhou Wen-ying attributed this pivotal role to the Liujia Channel in his project outline. In fact, after the Yuan Dynasty had replaced canal transport with maritime transport, the Liujia Channel had become the thoroughfare on which the tax rice and other kinds of goods of the Jiangsu and Zhejiang region were transported to the sea. In the early Ming, the maritime transport had not yet been abolished, and the Liujia Channel was still playing a decisive role.[54] When Zheng He (鄭和, 1371–1433) embarked for the first time on his famous voyage to the West in the third year of the Yongle reign (1405), he also started his journey from the Liujia Channel.[55] In other words, in the Yuan Dynasty and the Yongle era, the Liujia Channel was an unobstructed waterway, and it is thus unsurprising that Zhou Wenying singled it out. Furthermore, he made a rather forced statement when he said: "The Liujia Channel is the ancient Lou River (劉家港即古婁江)."[56] What he did was to misappropriate the definition for the "Three Rivers" given by Yu Zhongchu (庾仲初, ?–?) in the Jin Dynasty (1115–1124), the [Wu]song River, the Lou River and the Dong River, to support his own argument.[57]

Zhou Wenying's suggestion was not implemented in the Yuan Dynasty. But in his memorial, Xia Yuanji used exactly the same wording. He was obviously continuing to use this viewpoint without any reservation. Moreover, the reason why Xia Yuanji wanted to adopt an idea that had been created about a century earlier had to do with the fact that the Liujia Channel was kept deep and broad, due to the role it was playing in maritime transport in the Yuan Dynasty and the early Ming.

Things were very similar with the Baimao Channel. The Baimao Channel did not have a key position in Zhou Wenying's suggestion. It was only one of the many channels that should be dredged and connected to the river network in order to discharge the excess water in Kunshan (崑山), Jiading and Changshu (常熟) Counties.[58] The great turmoil at the end of the Yuan, however, made it a special waterway, such that it occupied an important place in the waterway network later. According to the *Gazetteer of Changshu County*:

> At the end of the Yuan, Zhang Shi-cheng seized Suzhou. As he read some old documents, he learnt about Zhou Wenying's suggestion. So he dredged the Baimao Channel, and extended it by over ninety *li*. It ran eastwards to the sea, its [course] was broad and straight, and its water was flowing in torrents. (元末, 張士誠據有蘇州. 閱故牘, 得周文英之議. 開塹白茆港, 長亙九十餘里, 東抵海口, 寬廣直逕, 水去如瀉.)[59]

We can see that Zhang Shicheng (1321–1367), who was one of the major figures at the end of the Yuan, drew inspiration from Zhou Wenying's suggestion. Hence he ordered the people to dig the Baimao Channel deep and broad, so it could be used to drain the water of Lake Tai smoothly into the Yangtze. This command, which was given in winter 1364, would have far-reaching repercussions.[60] As described in the "Memorials on the Conservancy of the Baimao Channel Baimao shuili shu (白茆水利疏)" of Yang Fang (楊舫, *juren* degree in 1477), "[T]he force of the current [of the Baimao Channel] in the upper reaches was great and inexhaustible (上流奔衝之勢大而不竭)." Consequently, "[T]he water of the channel was rapid, and its tides surged and ebbed in gurgling gushes (塘水湍急, 潮汐往來洶湧迅激)." So one reason why the bed of the Baimao Channel usually remained deep and broad was that it was abundant in water. Moreover, "[I]ts water flows forcefully, which helps keep it unobstructed (汩盪注洩, 益以通利)."[61] When Xia Yuanji came here forty years later to supervise hydraulic management, the Baimao Channel had been deepened by Zhang Shicheng, and therefore presented a convenient and solid route to drain off flood water. Under these circumstances, it was not unthinkable—it was even natural—for Xia Yuanji to take the unprecedented step of declaring the Baimao Channel as one of the Three Rivers.

The Liujia Channel and the Baimao Channel are both located at the upper reaches of the Wusong River, whereas the Huangpu River and the Fanjia Creek are on its lower reaches. And it was owing to Ye Zongxing's suggestion that these two waterways, the Huangpu and the Fanjia, made it into Xia Yuanji's plan. Research in recent years has shown that the lower reaches of the Wusong River silted up by the thirteenth century, and that they had become shallow and unnavigable. Therefore, gradually, the Huangpu River became the drain of the region.[62] But in the early years of the Ming Dynasty, even the Huangpu was silting. Consequently, Ye Zongxing submitted a memorial, in which "He asked to abandon the old [water]way, to dredge out the Fanjia Creek [instead], and to use it to guide the water of the [Huang]pu into the sea (請棄其故道, 濬范家浜,

引浦水以歸于海)." The Yongle Emperor was full of praise after he received this suggestion, and he commanded him to go and help Xia Yuanji.[63] Since the Yongle Emperor appreciated Ye Zongxing's suggestion so much, Xia Yuanji did not have any reason not to dredge the Fanjia Creek. So he connected it to the Wusong River, and, in this way, it became the last member of the Three Rivers.

Xia Yuanji presented this plan that revolved around the Three Rivers to the Yongle Emperor. It got approved, and so, in the winter of the same year, Xia requisitioned "over a hundred thousand men" to start working on the Liujia Channel.[64] After a brief return to court, he came here again to continue his incomplete work.[65] In the following year (1404), Xia Yuanji did not only dredge the Baimao Channel, the Huangpu River and the Fanjia Creek. Just like he had mentioned in his plan, he also drained all the channels on the southern and northern shores of the Wusong, such as the Anting (安亭), with the aim to manipulate them into the Yangtze or the sea.[66] So an argument about the Three Rivers, which was loaded with symbolical meaning and shaped by existing environmental factors, appeared in a specific historical situation—the Yongle Emperor's hope to legitimize his reign, and Xia Yuanji's urgent attempt to accomplish something great. And after it had been implemented in hydraulic engineering for two years in a row, it became the reality of the local river network. Or, to use the jargon of political ecology, it became a "fact" that was "taken for granted."

Reorganizing corvée: polder dykes and pedal pumps

In addition to dredging the Three Rivers, Xia Yuanji was also engaged in two other projects that were related to water control. And through these two projects, the Yongle regime displayed its efficiency in directing the mobilization of society, and it furthered the popular welfare, which would, again, be useful in consolidating the political legitimacy of the Yongle Emperor. First of all, Xia Yuanji supervised the regular construction of polder dykes. Shi Jian (史鑑, 1434–1496), a man from Wujiang (吳江) County, describes the situation at the time as follows:

> In the Yongle years, water management was conducted in the south-east. Great attention was paid to the dykes, which is something the Honourable Xia Zhongjin [i.e., Xia Yuanji] initiated in the beginning, and the commissioner of the Office of Transmission, Zhao Juren, continued later on. Both handpicked their officials, appointed them separately to the various counties, and the two honourable gentlemen observed and assessed them on a constant basis.
>
> The method they used was as follows: usually, they would call up labourers in early spring. Before each dyke was built, they would construct a model mound that should serve as a template and show how high and broad [the dyke] should be. And then, the dyke would be built accordingly. They took the earth from nearby fields, and, under their supervision, the people had to stabilise [the dykes] with sticks, so as to steady and consolidate them.

> After these works had been completed, they had the people fill up the fields, from which the earth had been taken, with bamboo baskets full of mud, until the [holes] were all replenished. Then they extended the basis inside and outside of the dykes, and called this "water blocker".... Furthermore, they instructed the people that they were allowed to plant indigo on the "water blocker," but that they were forbidden to plant beans. The reason was that indigo increases the soil, so that [the dyke] would get higher after a while, whereas beans erode the soil with their roots, so that [the dyke] would get lower after a while. (永樂中治水東南. 夏忠靖公創於前, 通政使趙居任繼於後, 無不注意於隄防. 皆妙選官屬, 分任諸縣, 而二公則周爰相度而考課焉.
>
> 其法, 常於春初編集民夫, 每圩先築樣墩以為式, 高廣各若干尺, 然後築隄如之. 其取土皆於附近之田, 又必督民以杵堅築, 務令牢固. 既訖工, 令民畚泥填灌取土之田, 必使充滿. 復於隄之內外, 增廣其基, 名為抵水 ... 又課民於抵水之上, 許其種藍而不許種豆, 蓋種藍則必增土, 久而日高, 種豆則土隨根去, 久而日低矣.)[67]

A few aspects of these measures that Xia Yuanji undertook deserve our attention. (1) Shi Jian said that Xia Yuanji was the first person to come up with the idea to construct model mounds for dyke building. This is a slight exaggeration, because a similar method had existed in the Yuan Dynasty.[68] But in the Ming Dynasty, it was indeed established for the first time by Xia Yuanji. (2) Xia Yuanji paid attention to quite a lot of details involved in the construction of polder dykes, including the creation of a model which the people could follow, which tools should be used, how the stability of the polder dykes could be guaranteed, which plants could (or could not) be planted, and so forth. (3) Since Xia Yuanji was intent on getting all the various details in the construction and maintenance of the dykes right, he took the choice and assessment of officials particularly seriously. As a result, the construction of the polder dykes did not only enhance flood prevention, but also increased the control over local officials. (4) Through handpicking his subordinate officials and assessing them more strictly, the ability of the government to organize and mobilize the people was increased as well. (5) As these were regular works, they became a constant display of the government's ability to assess its officials and to mobilize the people. In sum, the combination of these three elements—attention to details, control over officials and mobilization of the people—enhanced the local capacity for flood prevention.

In spite of all his hard work and well-planned efforts, however, Xia Yuanji was unable to fully protect the people from floods. From early June to early July in 1404, heavy rain inundated the low-lying fields of Suzhou Prefecture, where rice seedlings had only just been planted, with Wujiang County being affected worst.[69] The only thing Xia Yuanji could do in such a dire situation was to fall back on the "pedal pump rescue"—a method he had already employed the year before.[70] According to the "Sigh While Treading the Pump (Ta che tan 踏車歎)," which Xia Yuanji wrote when he was facing another flood in the third year of the Yongle reign, the "pedal pump rescue" worked like this:

> [I] carefully instruct the prefects and magistrates on how to reorganize [flood control]. Pedal pumps are mobilised and distributed around the fields. Alas! We have already gathered the pumps, but we have few people. So the farmers are called upon and checked [for their physical condition] and then sent to the villages. The sturdy and healthy women and men sign up their names. And we have them tread the pump, to pump off the stagnant flood water. They do not rest from dawn to dusk, and their feet make the pump wheels whirl as quickly as shooting stars. In the fields, the tax captains and community heads are sitting, beating the drum, and urging and rushing them not to slow down. (丁寧郡邑重規劃, 集車分布田周遭, 車兮既集人稀少, 點檢農夫下鄉保. 婦男壯健記姓名, 盡使踏車車宿潦, 自朝至暮無停時, 足行車轉如星馳. 糧頭里長坐擊鼓, 相催相迫惟嫌遲.)[71]

Japanese scholars have reached different conclusions on the workings and the nature of the “pedal pump rescue” in their analyses of the notion of *kyōdōtai* (共同體, community). Koyama Masaaki thinks that the tax captain and the community head were institutions through which the state imposed taxes and corvée onto society from top to bottom. Although they were in charge of local tasks, such as draining off water, they were not the spontaneous manifestation of an autonomous community.[72] Hamashima Atsutoshi holds that, in spite of the top-down nature of the “pedal pump rescue” system, it still had its root in the community because it actually relied on local leaders to mobilize the people. Such leaders were the tax captains and community heads, who existed autonomously in local society.[73] Okuzaki Hiroshi, on the other hand, thinks that this system certainly emerged spontaneously from society, but that it was not organized by the tax captains or community heads.[74]

If we carefully read a section by Mo Dan (莫旦, *juren* degree in 1465), to which all the three Japanese scholars refer in their arguments, we may gain a clearer understanding of the situation. According to Mo Dan:

> In Wu … the places [that are situated] near lakes and rivers are the lowest. Every [year], when the spring goes into the summer, rain falls incessantly. And then [the rivers and lakes] swell up and flow over. Inundations follow. The farmers [used to] gather the pedal pumps, which they called Great Chariot Pumps. Old and young, from far and near, they came together. And they usually beat the drums and clappers to determine when they might work and rest. There were times when they pumped in the morning and [the water] spilled over again in the evening so that they could not get a moment of rest over days and months.… When the Honourable Zhou Wenxiang [Zhou Chen] was grand coordinator, he commanded that each community head be entrusted with an Official Pump. If a town or a polder was inundated, tax captains would gather several Official Pumps, and supervise the labourers to collectively pump the water. [Consequently,] all the water was discharged in no time. Then the officials distributed grain rations as aid relief. (吳地 … 濱江傍湖最為低窪, 凡春夏之交, 梅雨連綿, 外漲泛溢,

> 渰沒隨之. 農家結集車戽, 號為大棚車. 人無老幼, 遠近畢集, 往往擊鼓鳴柝以限作息. 至有累日連月, 朝車暮漲而不得暫休者 ... 周文襄公巡撫之時, 令概縣排年里長, 每名置官車一輛, 假如某都某圍田被水渰沒, 則糧長拘集官車若干輛, 督令 人夫併工車戽. 須臾之間, 水去皆盡, 而又官給口糧以賑之.)[75]

Okuzaki Hiroshi has particularly pointed to the difference between the following two statements: "they pumped in the morning and [the water] spilled over again in the evening so that they could not get a moment of rest over days and months" and "all the water was discharged in no time." His explanation is that the original system was not as efficient as the one reorganized by the government. Furthermore, the tax captains and community heads did not have a place in this system until the government intervened in 1430 when Zhou Chen (周忱, 1381–1453) was sent to the Lower Yangtze Delta to restructure taxes and corvée. Okuzaki goes on saying that because the tax captains and the community heads had not played an important role in the spontaneously organised "pedal pump rescue" system, it was difficult, after Zhou Chen's departure, to maintain this Official Pump system.[76] To be sure, the most obvious contribution of Okuzaki's argument is to solve the question of why the "Official Pump" system ceased to exist. But if we look closely at the description Mo Dan gives prior to the aforementioned passage, we see that the original "pedal pump rescue" was quite similar to Okusaki's explanation. Mo only said, "the farmers [used to] gather the pedal pumps," which means that this was indeed a spontaneous action of the farmers. But he did not indicate that the "pedal pump rescue" existed because of the tax captain and community head's supervision and organization of the people. Neither did he imply that drums were played and clappers were beaten under the order of the tax captains or the community heads. In other words, the "pedal pump rescue" was certainly a spontaneous local system, but the tax captain and the community head did not have a prominent function in it.

The only shortcoming in Okuzaki's discussion is that he does not talk about Xia Yuanji's "Sigh While Treading the Pump." If we consider Mo's description and Xia's "Sigh While Treading the Pump" together, we find that, before Zhou Chen, Xia Yuanji had already made use of an existing pedal pump method to discharge the floodwater, and that he made two important changes to it: He gave the tax captains and community heads a special task, namely to "sit and beat the drum 坐擊鼓," and he changed the function of the drumbeat, from determining phases of work and rest, to being a tool that urged and rushed the treading of the pump. So the function of the tax captain and the community head emerged only when the government reformulated the structure of the "pedal pump rescue." Their role was not a constituent part of a spontaneous system.

Moreover, the "pedal pump rescue," in the version changed by Xia Yuanji, bears clear marks of official intervention. The prefects and magistrates were carefully instructed to improve the original method. When manpower was short, they had to call upon the farmers, check them for their physical condition, and then send them to the villages. In addition, they had to properly manage them by signing up their names.[77]

Xia Yuanji showed that he paid attention to the details in implementing both the "pedal pump rescue" and the "polder dyke construction," and he displayed how effective a control he had over local officials and the mobilization of the people. What was the political effect of these two methods? First of all, he increased the local capacity of emergency flood control. Second, in comparison with the dredging of the Three Rivers, these two methods advanced the reach of the government even closer to people's homes. And, finally, since the "polder dyke construction" was conducted regularly, people would feel even more often the presence of a capable government in their lives.

Replace hydraulic management with tax exemption and aid relief

In the roughly two years, from mid-1403 to mid-1405, Xia Yuanji dredged the Three Rivers and the many waterways which could serve as drains in the Lower Yangtze region, he came up with a method for "polder dyke construction," and he improved the system of "pedal pump rescue." However, in 1404 and 1405 floods were still inflicting heavy casualties in this region. And although the Yongle Emperor had been anxious to legitimize his reign through hydraulic management in 1403, he apparently had a different thought in the third year. This change of approach showed itself most clearly in the fact that he recalled Xia Yuanji to take charge of the Ministry of Revenues in the eighth month of that year.[78] The immediate cause for this was that Yu Xin, who had originally been in charge of the Ministry of Revenues, had died in this month.[79] But the fact that the Ministry of Revenues needed experienced and capable management was not the only reason.[80] Another one was that the Yongle Emperor had developed some doubts about water management. As mentioned above, Xia Yuanji had requisitioned over a hundred thousand people to dredge the Liujia Channel in 1403. Although there are no documents that would account for how many people he called up to dredge the Baimao Channel, the Fanjia Creek and the other channels in 1404, it should be reasonable to estimate that the total number was about the same as that for the Liujia Channel. So, in order to dig out the waterways, he had already mobilized about two hundred and fifty thousand people. According to the statistics of the twenty-sixth year of the Hongwu reign (1392), the Prefectures of Suzhou and Songjiang, which were situated in the hydraulic network of the Three Rivers, had a population of about 740,000 households altogether.[81] If we calculate that each household only dispatched one person, we see that, in only two years, one household in three had provided corvée labor for the hydraulic project. This was already an extremely heavy burden, compared with the taxes and corvée required for regular work within the *lijia* (里甲) system, as it had been established by the Hongwu Emperor.[82] If the Yongle Emperor continued proceeding along these lines, he would need to exploit people's labor even more. This would not be good for the establishment and maintenance of his political legitimacy. In other words, the method to win the people's hearts through hydraulic projects might, at that time, have already reached the point where it reversed itself and started producing the opposite effect.

The Yongle Emperor and Xia Yuanji were certainly alert to this negative effect. After the dredging of the Three Rivers had been completed, someone suggested to the Yongle Emperor "to call up the people to rent the land and cultivate (招民佃耕)" the fertile land after the flood water had retreated, as this would benefit the finances of the state. When Xia Yuanji received the edict from the Yongle Emperor, which instructed him to do so, he sighed: "The people are already utterly exhausted. It won't be enough to simply save them from death. But now [we] would like to impose even more labour upon them? (民疲已極! 赦死不暇, 况重役乎?)" Then he gave the Yongle Emperor a severe warning: "The pumping would be a fruitless exhaustion of the people's labour. And the season for planting has already passed [anyway]. How could this benefit the country? (車戽則徒勞民力, 栽種則已失時, 何益于國?)" This means that he regarded it as unwise at this time to bid the people again to pump off the remaining excess water, or to have the people plant grain. After the Yongle Emperor had received his memorial, he understood, and hence let the matter rest.[83] If the Yongle Emperor abandoned water pumping and land cultivation, there was even less reason to continue with the dredging of waterways, a labor which was even more strenuous.

In fact, it was not after Yu Xin's death that the Yongle Emperor decided to change his approach. In mid-1405 when the emperor once more sent out Xia Yuanji, Yu Shiji, Zhao Juren and Yuan Fu to deal with inundations in the Prefectures of Suzhou, Jiaxing, Songjiang and Huzhou (湖州), he ordered them not to manage the water but to bring aid relief. The Yongle Emperor said:

> The people of the four prefectures have suffered floods in consecutive years. Now the old grain is already used up, and the new seedlings have not yet grown. Old and young cry out, and they are starving helplessly. Can I and you, my ministers, simply enjoy our meals? Go and supervise the prefectures and counties, and quickly instruct the granaries to help them out. (四郡之民, 頻年厄於水患. 今舊穀已罄, 新苗未成, 老稚嗷嗷, 飢餒無告. 朕與卿等能獨飽乎? 其往督郡縣, 亟發倉廩賑之.)[84]

This was clearly a turning point. Previously, he had also granted aid relief to this region. But then aid relief had always run parallel to Xia Yuanji's supervision of hydraulic management.[85] Now these four officials, who had originally all belonged to the hydraulic management team, were assigned to direct the prefects and magistrates in opening the granaries for aid relief. Equally important is that he showed his sympathy for the hardship of the people by saying, "Can I and you, my ministers, simply enjoy our meals?" So aid relief was a means which could give him the image of a benevolent ruler without imposing the burden of corvée upon the people. And the Yongle Emperor started using this method to treat the people, who had been imperiled by the floods. He consented to the request put forward by Xia Yuanji to "distribute three hundred thousand *dan* of the stored grain for relief."[86] In the same year, he exempted the people who had been struck by floods in six prefectures, among them Suzhou and Songjiang, from taxes accounted for 3,379,700 *dan*.[87] In other words, although hydraulic management had originally seemed to be absolutely essential, it was now relinquished completely. This demonstrates that considerations of the political effects decided whether or not floods were handled through hydraulic strategies. In short, they were not a necessity.

After Xia Yuanji had returned to court, the hydraulic management in the Lower Yangtze Delta was conferred upon Zhao Juren. But apart from keeping the polder dykes in good condition, Zhao Juren did not accomplish anything else, such as dredging or digging waterways. From mid-1405, the Yongle Emperor apparently started responding to floods by granting aid relief or tax exemption. On this principle, he paid particular attention to those elements in Zhao Juren's administration that were related to this matter. For example, in 1412, the surveillance commissioner of Zhejiang claimed in a memorial that Zhao Juren had intentionally "concealed and not reported" that "due to heavy rainfall, the fields had not produced crops." The Yongle Emperor did not accept Xia Yuanji's objections that "the words of the surveillance commissioner were not to be fully trusted (按察之言, 未可悉信)." He sent out an official to make investigations. "If the inundations had destroyed the seedlings, [the region] should be exempted from taxes. If the people were [afflicted] even more seriously with the floods, the officials should distribute grain to help them. (但苗壞於水者, 蠲其稅, 民被水甚者, 官發粟賑之)."[88] This shows that the Yongle Emperor understood very well Zhao's habit to fawn upon the emperor while treating the people badly. For the Yongle Emperor, who was hoping to maintain the image of a benevolent ruler, this was not something to be encouraged. So he overruled Zhao Juren's reports several times, and granted the people who had suffered from floods ample and generous tax exemptions and aid relief.

Equally noteworthy is that the Yongle Emperor never mentioned hydraulic management in these directives. This absence was in line with other actions he undertook in this period. For instance, in 1412, because of the floods, "[He] exempted the Counties of Wujiang, Changzhou (長洲), Kunshan and Changshu in [South] Zhili from paying grain taxes in the amount of about 138,690 *dan*."[89] In 1413, he exempted the Counties of Changzhou and Kunshan, which had been afflicted by inundations, from paying the autumn grain tax.[90] In 1414, in addition to the help he gave the people who were starving because of floods in Wuxian and Changshu County,[91] he also completely did away with taxes for the Prefectures of Suzhou, Songjiang, Jiaxing, Huzhou and Hangzhou (杭州).[92] This demonstrates very clearly that there were two reasons why the Yongle Emperor abandoned hydraulic management. First, the Ministry of Revenues was in urgent need of a superior official at its top; and second, he judged that excessive deployment of people's labor would harm his political legitimacy. As a result, tax exemption and emergency relief became his chief strategies in dealing with the victims of natural disasters.

Conclusion

Modern scholars find Xia Yuanji's water management, which revolved round the "Three Rivers," very problematic. The reason is that it did not dredge the whole Wusong River, which aggravated its siltation. Moreover, the alternative waterways did not have enough capacity to drain off all the water. As a result, later generations were confronted with an even more complicated situation.[93]

This view is doubtlessly true if we look at the consequences for later times. But, as I have shown in this chapter, Xia Yuanji's decision emerged out of his specific historical circumstances. And these specific historical circumstances consisted of numerous interacting factors, which can only be understood through an in-depth examination. This chapter not only attempts to gain a sympathetic understanding of historical figures and their actions, but also hopes to make a methodological suggestion: it would be fruitful if the study of hydraulic history explored the political expectations, cultural implications, the limits and possibilities inherent in existing environmental conditions, and the combined effects of all these factors. This would give us a deeper understanding of the cross-sections of history.

Furthermore, drawing upon the research presented in this chapter, I should point out here that Xia Yuanji's measures, which revolved around the Three Rivers, had a considerable normative function in the discourse and practice of Ming water. For instance, when the Prefect of Suzhou, Kuang Zhong (況鍾, 1384–1442), submitted a memorial in the ninth year of the Xuande (宣德) reign (1434), in which he asked to create the position of an official who would be exclusively in charge of water management, he referred to the Wusong River, the Liujia Channel and the Baimao Channel as the Three Rivers, and he built his argument upon this definition.[94] In the seventh year of the Hongzhi (弘治) reign (1494), Qin Qing (秦慶, ?–?) asked to send an official who would supervise hydraulic management. Although he did not clarify what the Three Rivers were, he still used this expression as a key term to persuade the court.[95] Even at the end of the Ming, when Zhang Guowei (張國維, 1595–1646) was tracing the "course of the rivers (水脈)" to compile his *Complete Record of Hydraulic Management in the Wu Region* [吳中水利全書], he still had to define what the Three Rivers were in his time.[96] And, his description fits very well with the plan Xia Yuanji had presented two centuries earlier. And those actually carrying out hydraulic management had to give some thought and consideration to the Three Rivers as well. For instance, when assisting Li Chongsi (李充嗣, 1462–1528) with water management in the Lower Yangtze Region in 1522, Yan Ruhuan (顏如環, ?–?) was so determined to dredge the Wusong River that he asserted with very powerful words that it would not be satisfactory to force the water of the Wusong River into the Liujia Channel.[97] Although Hai Rui (海瑞, 1514–1587) first dredged the Wusong River in the Longqing (隆慶) period (1567–1572), the Three Rivers were still the basic framework for his understanding of the region. Therefore, after he had dredged the Wusong River, he requested in a memorial to continue with digging out the Baimao Channel. The idea was that it would back up the drainage system of the eastern part of Lake Tai, together with the Liujia Channel, which was still unobstructed at the time.[98]

The above examples clearly show that although the Three Rivers had been a hydraulic strategy employed by Xia Yuanji under the specific historical circumstances of the early Ming, they became the blueprint which later generations followed, and with which even those who disagreed had to consider carefully. In other words, they laid out the possibilities and limits of the policies and actions of later generations, because they were "taken for granted" as precedents. The way later generations responded, in their discourses and practical actions, to the norms created by earlier generations would be a topic worth exploring in the future.

Notes

1 This translation is taken from Peter Ditmanson, "Venerating the Martyrs of the 1402 Usurpation: History and Memory in the Mid and Late Ming," *T'oung Pao* 93 (2007): 111.

2 Benjamin A. Elman, "The Formation of 'Dao Learning' as Imperial Ideology During the Early Ming Dynasty," in Theodore Huters et al. (eds.), *Culture and State in Chinese History: Conventions, Accommodations, and Critiques* (Stanford: Stanford University Press, 1997), pp. 58–82.

3 Hok-Lam Chan, "Legitimating Usurpation: Historical Revisions under the Ming Youngle Emperor (r. 1402–1424)," in Philip Yuen-sang Leung (ed.), *The Legitimation of New Orders: Case Studies in World History* (Hong Kong: Chinese University Press, 2007), pp. 75–158. Chan covers more ground in his article than indicated here. He also analyzes how later reigns received these documents which the Yongle Emperor had rewritten, and how, through this process, the Jianwen Emperor regained his place in the history of the dynasty. For related discussions see also Ditmanson, "Venerating the Martyrs of the 1402 Usurpation," pp. 110–158.

4 Zhang Xiaoping 張小平 and Zhou Shaoqing 周少青, "Ming Chengzu in the Field of Political Legitimacy—An Exemplary Reading (政治合法性視野中的明成祖一個範本的解讀)," *Jilin Normal University Journal* (*Humanities & Social Science Edition*) [吉林師范大學學報 (人文社會科學版)] No. 2 (2008): 59–64.

5 Cf. Zhu Hong 朱鴻, *Ming Chengzu and Yongle Politics Ming* [明成祖與永樂政治], (Taipei: Guoli Taiwan shifan daxue lishi yanjiusuo, 1988), pp. 190–191; Li Chaoran 李焯然 and Mao Peiqi 毛佩琦, *Ming Chengzu's Theory of History* [明成祖史論], (Taipei: Wenjin chubanshe, 1994), pp. 66–88; Chao Zhongchen 晁中辰, *Biography* of *Ming Chengzu* [明成祖傳], (Taipei: Taiwan Shangwu yinshuguan, 1996), pp. 489–499; Cai Taibin 蔡泰彬 has discussed this topic from the perspective of Xia Yuanji himself. See Cai Taibin, "The Elder Statesman of Five Dynasties, Xia Yuanji, and His Water Control in the Lower Yangzi Delta (明代五朝元老夏原吉治水江南)," *Journal of Paisa Humanities and Social Sciences* [白沙人文社會學報] No. 1 (2002): 205–226.

6 See the series of articles by Wang Jiange 王建革 and Xie Shi 謝湜 in the bibliography.

7 Wang Jiange 王建革, "From the Mouth of the Three Rivers to the Three Rivers—A Forced Interpretation of Lou and Dong River and its Influence (從三江口到三江—婁江與東江的附會及其影響)," *Social Science Research* [社會科學研究] No. 5 (2007): 162–168.

8 There are four reviews on this book. For details please see the bibliography at the end of this chapter.

9 Paul Robbins, *Political Ecology: A Critical Introduction* (Malden, MA: Blackwell Publishing, 2004), p. 85.

10 Ibid., p. 108.

11 Ibid., pp. 109–110.

12 Yang Shiqi 楊士奇 (1364–1444) et al., *Veritable Records of Ming Taizong* [明太宗實錄], (Kyoto: Chûbun shuppansha, 1984), 9b: entry for the 己巳 day of the 6th month of Jianwen 4 (17 July 1402).

13 Ibid., 10b: entry for the 丙午 day of the 7th month of Hongwu 35 (23 August 1402).

14 Ibid., 12a: entry for the 甲申 of the 9th month of Hongwu 35 (30 September 1402).

15 Ibid., 19: entry for the 己酉 of the 4th month of Yongle 1 (23 April 1403).

16 Tan Qian 談遷 (1594–1657), *Discussions about the States* [國榷], (Shanghai: Shanghai guji chubanshe, 1996), 12: 1–2. Chen Jian 陳建 (1497–1567), *The Essential Collection of the Comprehensive Imperial Records* [皇明通紀集要] (Beijing: Beijing chubanshe, 1995), 12: 17, gives a list of different people, which also counts one person more.

17 They all died in July 1402, cf. Yang Shiqi, *Veritable Records of Ming Taizong*, 9b: entry for the 丁丑 day of the 6th month of Jianwen 4 (25 July 1402).

18 Lian Zining and Fang Xiaoru died on the same day. Cf. Tu Shu-fang 屠叔方 (*jinshi* 進士 degree in 1577), *Government and Public in the Jianwen Reign* [建文朝野彙編], (Taipei: Zhuangyan wenhua shiye youxian gongsi, 1996), 10: 31. For the dates when other men were executed or committed suicide, see Liu Tingluan 劉廷鑾 (*gongsheng* 貢生 degree in 1662), *A Monthly Chronology of the Abdication of the Jianwen Emperor* [建文遜國之際月表], (Taipei: Yiwen yinshuguan, 1970), pp. 27–34.

19 Huang Zuo 黃佐 (1490–1566), *Eradicating Things Past* [革除遺事], (Taipei: Mingwen shuju, 1991), 6: 9. Please note that the list Yongle refers to here only mentions twenty-nine people. This differs from the records in the *Discussions about the States.* And since the *Eradicating Things Past* does not elaborate who these twenty-nine people were, whereas the *Discussions about the States* explicitly lists all of them, I follow the *Discussions about the States* in this chapter.

20 I will analyze this method Yongle employed more closely below. Let me only make one point here: It did not guarantee Jianwen's former ministers a secure position in the new reign if they had surrendered to Yongle and urged him to seize power after his occupation of Nanjing. Following the accounts in the *Discussions about the States*, Kawagoe Yasuhiro 川越泰博 has described how people who had done so were put into important positions in the Yongle reign, cf. Kawagoe Yasuhiro, *History of the Ming Dynasty's Jianwen Reign* [明代建文朝史の研究], (Tōkyō: Kyūko shoin, 1997), pp. 435–442. But if we consider that the Yongle Emperor threatened Xie Jin, it becomes clear that the old Jianwen ministers could not earn trust in the new reign by simply having surrendered and urged the Yongle Emperor to seize power. They had to make other contributions.

21 Xia Chongwen 夏崇文 (1456–1507), *The Deeds of Xia Zhongjing* [夏忠靖公遺事], (Taipei: Zhuangyan wenhua shiye youxian gongsi), p. 5. In the first year of the Jianwen reign, Xia Yuanji moved from Fujian to Qizhou. I am not aware of any documents according to which he would have moved again to another place or returned to Nanjing immediately before Yongle ascended to the throne. Therefore I will assume for the moment that he was still in Qizhou. The Compilers of the *Siku Quanshu* 四庫全書 [*Complete Library in Four Sections*] believe, the story that Xia Yuanji pledged allegiance to Yongle because someone had "arrested him to present him to the Prince of Yan" was invented by Xia Yuanji's children and grandchildren, to cover up for him. See Yong Rong 永瑢 (1744–1790), *Index to the Siku Quanshu* [欽定四庫全書總目], (Taipei: Taipei Shangwu yinshuguan, 1983), 16: 11. But these Compilers fail to explain the reason why they say this. On the contrary, from what they write in their text overall, it seems as if ideals of loyalty and morality prompt them to this judgment.

22 Zhang Qin 張芹 (1466–1541), *A Record That Preserves What Might Be Forgotten* [備遺錄] (Taipei: Mingwen shuju, 1991), p. 16.

23 Cao Shenfang 曹參芳, *Uprightness during the Abdication* [遜國正氣紀], (Taipei: Mingwen shuju, 1991), pp. 609–612; Chen Jian, *Comprehensive Imperial Records and Laws* [皇明通紀法傳全錄], (Beijing: Beijing chubanshe, 2005), 13: 3.

24 Tang Hezheng 唐鶴徵 (*jinshi* degree in 1571), *Imperial Ministers* [皇明輔世編], (Taipei: Mingwen shuju, 1991), p. 150; Yin Shouheng 尹守衡 (1551–1633), *A Secret History of the Ming* [皇明史竊], (Shanghai: Shanghai guji chubanshe, 1996), 3: 8.

25 Yang Shiqi, *Veritable Records of Ming Taizong*, 11: entry for the 丁巳 day of the 8th month of Hongwu 35 (3 September 1402).

26 The Yongle Emperor adopted Xia's advice on leniently dealing with salt merchants. See Yang Shiqi, *Veritable Records of Ming Taizong*, 14: entry for the 壬午 day of the 11th month of Hongwu 35 (27 November 1402), and 15: entry for the 乙丑 day of the 12th month of Hongwu 35 (9 January 1403). When tackling the explosion of rice prices in Henan and Shandong, Xia Yuanji suggested to adjust the amount of rice that could be bought with one *yin* of salt. In addition to these two suggestions that concerned the equitable exchange of rice for salt, the Yongle Emperor also agreed with Xia Yuanji's proposal to adjust the percentage of the officials' salary that would be paid in cash. Cf. Yang Shi-qi, *Veritable Records of Ming Taizong*, 15: entry for the 甲寅 day of the 12th month of Hongwu 35 (29 December 1402).

27 Yang Shiqi, *Veritable Records of Ming Taizong*, 12b: entry for the 癸卯 day of the 9th month of Hongwu 35 (19 October 1402).

28 Ibid., 14: entry for the 己亥 day of the 11th month of Hongwu 35 (14 December 1402).

29 There are other similar incidents. See Yang Shi-qi, *Veritable Records of Ming Taizong*, 17: entry for the 丁卯 day of the 2nd month of Yongle 1 (12 March 1403); *Veritable Records of Ming Taizong*, 1: entry for the 己酉 day of the 4th month of Yongle 1 (23 April 1403); Huang Guangsheng 黃光昇 (1506–1586), *Rules of an Enlightened Reign* [昭代典則], (Taipei: Zhuangyan wenhua shiye youxian gongsi, 1996), 13: 2.

30 According to *The Illustrated Records of Jiaxing Prefecture* [嘉興府圖記], compiled by Zhao Ying 趙瀛 (*jinshi* degree in 1529) and Zhao Wen-hua 趙文華 (*jinshi* degree in 1529) in the Jiajing reign (1522–1566), no floods are recorded for the Hongwu, Jianwen and Yongle reigns. Cf. *The Illustrated Records of Jiaxing Prefecture* [嘉興府圖記], (Taipei: Zhuangyan wenhua shiye youxian gongsi, 1996), 20: 6. According to the *Gazetteer of Suzhou Gusu zhi* [姑蘇志], written by Wang Ao 王鏊 (1450–1524) in the Zhengde reign (1506–1521), no floods happened after Hongwu 9 (1376) up until the second year of the Yongle reign. Cf. *Gazetteer of Suzhou Gusu zhi* [姑蘇志], (Taipei: Shangwu yinshuguan, 1983), 12: 28. *The Gazetteer of Songjiang Prefecture* [松江府志] compiled by Chen Wei 陳威 (*jinshi* degree in 1490) and Gu Qing 顧清 (1460–1528) recorded that there were no floods after Hongwu 3 (1370). Only in the "first years of the Yongle reign," people suffered again from "floods several years in a row 連歲大水." Cf. *The Gazetteer of Songjiang Prefecture* (*Zhengde Edition*) [(正德)松江府志], (Taipei: Zhuangyan wenhua shiye youxian gongsi, 1996), 32: 25.

31 Shen Qi 沈棨 (1501–1568) recorded in the *Investigation of the Waters in Wujiang County* [吳江水考], (Taipei: Zhuangyan wenhua shiye youxian gongsi, 1996), 2: 14–15, "In the thirty years of the Hongwu reign of Taizu, there were no natural disasters (太祖洪武共三十年間俱無災)." The first time floods were recorded in the Ming Dynasty was in the second year of the Yongle reign. Zhang Neiyun 張內蘊 and Zhou Dashao 周大韶, who helped Lin Yingxun 林應訓 (*jinshi* degree in 1571) with hydraulic management in the first years of the Wanli reign (1573), recorded the same in their book *The Waters of the Wu, Wuxing, and Huiji* [三吳水考], 6: 5. The only account which shows that there were floods in the Hongwu reign is the *Complete Record of Hydraulic Management in the Wu Region* [吳中水利全書] by Zhang

Guowei 張國維 (1594–1645), who undertook water management in the Lower Yangzi Delta in the Chongzhen 崇禎 reign (1628–1644). This book mentions that there were "inundations" ("大水") in the first (1368), seventh (1374) and ninth year (1376) of the Hongwu reign. And in the twenty-third year (1390), the water of the sea overflew the shores and "drowned the farmland of the people (漂沒民田)." Afterwards, it states as well that the next flood only happened in the second year of the Yongle reign. Cf. *Complete Record of Hydraulic Management in the Wu Region*, (Taipei: Shangwu yinshuguan, 1983), 8: 11–12.

32 Cf. Li Jinglong 李景隆, *Veritable Records of Ming Taizu* [明太祖實錄], (Kyoto: Chûbun shuppansha, 1984), 234: entry for the 乙亥 day of the 8th month of Hongwu 27 (3 September 1394). For a discussion on how the Hongwu Emperor dispatched the students of the Imperial Academy to undertake hydraulic management, see Lin Liyue 林麗月, *The Students of the Imperial Academy in the Ming Dynasty* [明代的國子監生], (Taipei: Sili Dongwu daxue Zhongguo xueshu zhuzuo jiangzhu wenyuanhui, 1978), pp. 26–31.

33 Li Jinglong, *Veritable Records of Ming Taizu*, 15: entry for the 丁丑 day of the 12th month of Hongwu 35 (21 January 1403).

34 Cf. Han Jun 韓浚 (*jinshi* degree in 1598), *Gazetteer of Jiading County* (*Wanli Edition*) [(萬曆) 嘉定縣志], (Taipei: Zhuangyan wenhua shiye youxian gongsi, 1996), 14: 1–14; 17: 13–14. In 1605, when Han Jun (fl. 1600–1620s) compiled this book of 20 *juan*, he was the magistrate of Jiading County.

35 Among all these books on water conservancy and local gazetteers mentioned above, only *Investigation of the Waters in Wujiang County* by Shen Qi is a private compilation. But as Shen Qi was once secretary in the Work Project Office of the Ministry of Works, he had the opportunity to access government files for writing this 5 *juan*-book on water management. So, even though the *Investigation of the Waters in Wujiang County* is of a more private nature, its data are still based on official files.

36 Yang Shiqi, *Veritable Records of Ming Taizong*, 19: entry for the 己酉 day of the 4th month of Yongle 1 (23 April 1403).

37 Ibid., 22: entry for the 辛亥 day of the 6th month of Yongle 1 (24 June 1403).

38 Ibid., *Veritable Records of Ming Taizongu*, 22: entry for the 戊申 day of the 8th month of Yongle 1 (20 August 1403). Unfortunately, the historical material available today does not contain any details about the *Shuili ji*, which Yu Shiji was ordered to give to Xia Yuanji.

39 Yang Shiqi, *Veritable Records of Ming Taizong*, 19: entry for the 乙巳 day of the 1st month of Yongle 2 (13 February 1404).

40 Zhang Tingyu 張廷玉 (1672–1755) et al. (ed.), *History of the Ming* [明史], (Taipei: Zhonghua shuju, 1965), 149: 4150.

41 Zhang Guowei, *Complete Record of Hydraulic Management in the Wu Region*, 9: 6–7. In the original text, Zhang Guo-wei mistakes Yao Bo-shan for "Yao Shan." The loyal minister of Jianwen, Yao Shan, had already died in the fourth year of the Jianwen reign (1402). But Yao Bo-shan was from Changshu, and had taken his civil service examinations in 1399, the first year of the Jianwen reign. He did indeed take part in Xia Yuanji's work, in the function of supervising secretary of the Office of Scrutiny for Works. Cf. Feng Rubi 馮汝弼 (1499–1577) and Deng Fu 鄧韍 (*juren* degree in 1516), *The Gazetteer of Changshu* (*Jiajing Edition*) [(嘉靖) 常熟縣志], (Beijing: Shumu wenxian chubanshe, 1988), 3: 24.

42 Chen Wei and Gu Qing, *The Gazetteer of Songjiang Prefecture* (*Zhengde edition*), 29: 9; 1: 11.

43 Yang Shiqi, *Veritable Records of Ming Taizong*, 22: entry for the 戊申 day of the 8th month of Yongle 1 (20 August 1403).

44 Jiao Hong 焦竑 (1541–1620), *Records of the Testimonies about the Worthies of the Reigning (Ming) Dynasty* [國朝獻徵錄], (Taipei: Mingwen shuju, 1991), 69: 1.

45 Zhang Guowei, *Complete Record of Hydraulic Management in the Wu Region*, 27: 2. In this text Xia used historical allusions to assert the special position of the Yongle Emperor, and to tie in his hydraulic work with the legitimization of the Yongle reign. Significantly, this text echoed Yongle's plan very suitably. Apparently, Xia Yuanji had learnt his lesson from the experience he had had in the previous year, when his proposal for carving out new woodblocks to print paper money had been rejected by the emperor. And he understood now that his hydraulic management had a political implication, which was tied up with the legitimization of the reign.

46 Xia Yuanji was extremely anxious to attribute his merits to the Yongle Emperor. There is a story that shows vividly how much care he proceeded with. See Chen Jian, *Comprehensive Imperial Records and Laws* [皇明通紀法傳全錄], (Beijing: Beijing chubanshe, 2005), 14: 22.

47 Yang Shiqi, *Veritable Records of Ming Taizong*, 22: entry for the 戊申 day of the 8th month of Yongle 1 (20 August 1403).

48 See, for example, Fu Yin 傅寅, *Judgements about the Tribute of Yu* [禹貢說斷], (Taipei: Shangwu yinshuguan, 1983), 2: 14–17; and Mao Huang 毛晃, *Guide to the Tribute of Yu* [禹貢指南], (Taipei: Shangwu yinshuguan, 1983), 1: 21–26.

49 For further details on the frequent criticism which Cai Shen's annotations have met with since the Song Dynasty, and on the principles and the process of creating the *Collected Selection of the Commentaries of the Book of Documents*, see Yong Rong, *Index to the Siku Quanshu*, 12: 11–14.

50 Liu Sanwu, *Collected Selection of the Commentaries of the Book of Documents*, (Taipei: Taibei shangwu yinshuguan, 1983), 2: 10.

51 Li Jingde 黎靖德 (fl. 1270s) (ed.), *Master Zhu's Dialogues* [朱子語類], (Taipei: Huashi chubanshe, 1987), 2025.

52 From the perspective of political ecology, the "Three Rivers" did not only disallow other options, but also restricted Xia Yuanji's possibilities in that he had to pick three, no fewer and no more, main channels from the many waterways. This also shaped the way in which the resources were mobilized and used.

53 Gui Youguang 歸有光 (1507–1571), *Records of Water Conservancy in the Three Wu* [三吳水利錄], 3: 4–6. There is no direct evidence that could explain how big ships of three- or five-thousand *liao* were. But Wu Zimu 吳自牧 (1161–1237) says in his *A Dream of Millet* [夢粱錄], (Taipei: Shangwu yinshuguan, 1983), 12: "The vessels of sea merchants are of varying sizes. The big ones are 5,000 *liao*, and they can carry five- or six-hundred men (海商之艦, 大小不等, 大者五千料, 可載五六百人)." This indicates that the ships riding at anchor in the Nanshiqiao Channel were quite big.

54 Qiu Jun 丘濬 (1421–1495), *Supplement to the Extended Meanings of the Great Learning* [大學衍義補], (Taipei: Shangwu yinshuguan, 1983), 34: 8–9.

55 For related discussions, see Mao Boke 茅伯科 and Tian Nanfan 田南帆, "The Liujia Channel at the Time of Zheng He's Voyage (鄭和下西洋時期的劉家港)," *Journal of Shanghai University* (*Social Sciences Edition*) [上海大學學報 (社會科學版)] No. 2 (1985): 71–76; Wu Yuming 吳聿明, "Zheng He's Voyage and Shipwrecks in Taicang in the Yuan and Ming (鄭和下西洋與太倉元明沈船之研究)," *Shanghai Shipbuilding* [上海造船] No. 2 (2005): 40–42.

56 Gui Youguang, *Records of Water Conservancy in the Three Wu*, 3: 7.

57 Cf. Mao Huang, *Guide to the Tribute of Yu*, 1: 21. One possible explanation why Zhou Wenying chose Yu Zhongchu's definition of the "Three Rivers," and not any other one of the many available interpretations, is that Yu's definition contained the "Lou River," and that "*Loujiang*" ("Lou River") sounds similar to "Liujia Channel." So it is rather unsurprising that he misappropriated it to make his point. Moreover, a search of the *Wenyuange siku quanshu* 文淵閣四庫全書, the *Zhongguo jiben guji ku* 中國基本古籍庫 and the *Zhongguo difangzhi ku* 中國地方志庫 for hints at when these two names appear together has also shown that Zhou Wenying was the first one to hold this view, and Xia Yuanji was the first one to put this view into practice. After

the early Ming, these two names appeared together very often in the context of the "Three Rivers." So the fact that the Liujia Channel was promoted to one of the Three Rivers was a historical product, created and consolidated by Zhou Wenying and Xia Yuanji together.

58 Gui Youguang, *Records of Water Conservancy in the Three Wu*, 3: 6.

59 Sang Yu 桑瑜 (*juren* degree in 1459), *Gazetteer of Changshu* (*Hongzhi Edition*) [(弘治) 常熟志], (Taipei: Zhuangyan wenhua shiye youxian gongsi, 1996), 1: 45.

60 For the dating of these events, see Qian Qianyi 錢謙益 (1582–1664), *Biographies of Major Figures at the Beginning of this Dynasty* [國初群雄事略], (Beijing: Beijing chubanshe, 1997), 7: 52.

61 Yang Yi 楊儀 (*jinshi* degree in 1526), *Literature from the Ancient Wu* [古虞文錄], (Taipei: Zhuangyan wenhua shiye youxian gongsi, 1996), 1: 40.

62 Cf. Wang Ting 王頲, "The Management of the Wusong River in the Yuan Dynasty, and the Problem of the Change of Course of the Main Waterway (元代的吳淞江治理及幹道「改流」問題)," *Collections of Essays on Chinese Historical Geography* [中國歷史地理論叢] 18: 4 (2003): 83–91; Man Zhimin 滿志敏, "Causes of the Formation of the River Network of the Huangpu River (黃浦江水系形成原因述要)," *Fudan Journal* (*Social Sciences Edition*) [復旦學報 (社會科學版)] No. 6 (1997): 54–59; Fu Linxiang 傅林祥, "Wusongjiang xiayou yanbian xin jie 吳淞江下游演變新解 (A New Explanation of the Development of the Lower Reaches of the Wusong River)," *Academic Monthly* [學術月刊] No. 8 (1998): 89–94.

63 Chen Wei and Gu Qing, *The Gazetteer of Songjiang Prefecture* (*Zhengde edition*), 29: 9.

64 For details on the number of labourers he requisitioned in the first year of the Yongle reign, cf. Xu Xueju 徐學聚 (*jinshi* degree in 1583), *Collection of the Canons of Our Dynasty* [國朝典彙], (Taipei: Zhuangyan wenhua shiye youxian gongsi, 1996), 191: 4. In this year he apparently only dealt with the Liujia Channel. Cf. Qian Sule 錢肅樂 (1606–1648) and Zhang Cai 張采 (1596–1648), *The Gazetteer of Taicang Subprefecture* (*Chongzhen Edition*) [(崇禎)太倉州志], (Edition of the 17th Year of the Kangxi Reign [1678–1679]), 7: 12.

65 Yang Shiqi, *Veritable Records of Ming Taizong*, 27: entry for the 乙巳 day of the 1st month of Yongle 2 (13 February 1404).

66 For related records, see the following texts: Yang Shiqi, *Veritable Records of Ming Taizong*, 24: entry for the 戊辰 day of the 9th month of Yongle 2 (2 November 1404); Qian Su-le and Zhang Cai, *The Gazetteer of Taicang Subprefecture* (*Chongzhen Edition*), 7: 12.

67 For related records, see the following texts: Yang Shiqi, *Veritable Records of Ming Taizong*, 24: entry for the 戊辰 day of the 9th month of Yongle 2 (2 November 1404); Qian Su-le and Zhang Cai, *The Gazetteer of Taicang Subprefecture* (*Chongzhen Edition*), 7: 12.

68 For information on the "five standards of model polder dykes (五等圍岸體式)" in the Yuan Dynasty, see Yao Wenhao 姚文灝 (1455–1504), *Documents of Water Management in Zhexi* [浙西水利書], (Taipei: Shangwu yinshuguan, 1983), 2: 20.

69 Mo Dan 莫旦, *The Gazetteer of Wujiang* (*Hongzhi Edition*) [(弘治) 吳江志], (Taipei: Xuesheng shuju, 1987), 12: 16.

70 I argue that Xia Yuanji had already used this system in the first year of the Yongle reign because Mo Dan says in the same section: "In the previous year, the farmer families in high-lying villages were sent to pump out the low-lying fields (高鄉田家, 去年被差車低田)."

71 Chen Wei and Gu Qing, *The Gazetteer of Songjiang Prefecture* (*Zhengde edition*), 32: 25.

72 Koyama Masaaki 小山正明, *A Study of Economic and Social History in the Ming and Qing* [明清社会経済史研究], (Tōkyō: Tōkyō daigaku shuppan kai, 1992), pp. 222–224.

73 Hamashima Atsutoshi 濱島敦俊, *A Study of the Rural Society in the Lower Yangzi Region in the Ming Dynasty* [明代江南農村社会の研究], (Tōkyō: Tōkyō daigaku shuppan kai, 1982), pp. 9–18.
74 Okuzaki Hiroshi 奥崎裕司, *A Study of Chinese Gentry Landlords* [中国郷紳地主の研究], (Tōkyō: Kyūko shoin, 1978), pp. 528–530.
75 Mo Dan, (*Hongzhi*) *Wujiangzhi*, 5: 23–24.
76 On the achievements of Zhou Chen, see Yu Weiming 郁維明, *The Economic and Social Reforms in the Lower Yangzi Region Undertaken by Zhou Chen in the Ming Dynasty* (in Chinese), (Taipei: Shangwu yinshugan, 1990).
77 Mo Dan, *The Gazetteer of Wujiang* (*Hongzhi Edition*), 12: 16.
78 Lei Li 雷禮 (1505–1581), *The Great Officials of the Dynasty* [國朝列卿紀], (Taipei: Mingwen shuju, 1991), 6: 51.
79 Yang Shiqi, *Veritable Records of Ming Taizong*, 45: entry for the 戊辰 day of the 8th month of Yongle 3 (29 August 1405), and Yin Shouheng, *A Secret History of the Ming*, 43: 2.
80 By now, the Yongle Emperor already had a firm grasp of his regime's authority, because before mid-1405, the efforts the Yongle Emperor had undertaken to legitimize his reign had progressed considerably in other areas. As far as foreign politics were concerned, Korea, the Ryukyu Islands and Vietnam had come to court to congratulate him not long after he had ascended to the throne. And in late 1403, Japan sent a tribute commission. Cf. Yang Shiqi, *Veritable Records of Ming Taizong*, 45: entry for the 己亥 day of the 9th month of Yongle 1 (10 October 1403). With respect to domestic politics, he had already done the first step of his scheme—which was to rewrite the historical record in a way that it claimed the "quelling of disturbances" had been unavoidable—when the *Vertical Records of Ming Taizu* were completed for the first time in mid-1403. Cf. ibid., 21: entry for the 辛酉 day of the 6th month of Yongle 1 (4 July 1403). He made progress in winning the scholars to his side, when the first civil service examinations were held in mid-1404.
81 Shen Shixing 申時行 (1535–1614), *Laws and Institutions in the Ming* [明會典], (Beijing: Zhonghua shuju, 1989), 19: 124.
82 The digging of waterways was not part of the *lijia* corvée system. See Liang Fangzhong 梁方仲, *Collection of Liang Fangzhong's Works on the History of Economics* [梁方仲經濟史論文集], (Beijing: Zhonghua shuju, 1989), p. 40; p. 588. This sort of corvée was summoned as occasion demanded, and the work was sometimes harder, sometimes lighter. But if we compare the demands in these dredging projects with the normal rotation frequency in the *lijia* corvée system, we can estimate that these projects, which lasted for two consecutive years, had a great impact on people's daily lives in the region.
83 Yang Shiqi, *Veritable Records of Ming Taizong*, 43: entry for the 甲申 day of the 6th month of Yongle 3 (16 July 1405).
84 Ibid.
85 Ibid., 19: entry for the 己未 day of the 4th month of Yongle 1 (3 May 1403); 21: entry for the 乙未 day of the 7th month of Yongle 1 (7 August 1403); 32: entry for the 乙未 day of the 6th month of Yongle 2 (1 August 1404).
86 Jiao Hong, *Records of the Testimonies about the Worthies of the Reigning (Ming) Dynasty*, 28: 7.
87 Yang Shiqi, *Veritable Records of Ming Taizong*, 46: entry for the 丁酉 day of the 9th month of Yongle 3 (27 September 1405).
88 Ibid., 129: entry for the 壬申 day of the 6th month of Yongle 10 (27 July 1412).
89 Ibid., 130: entry for the 丙戌 day of the 7th month of Yongle 10 (10 August 1412).
90 Ibid., 141: entry for the 乙巳 day of the 7th month of Yongle 11 (24 August 1413).
91 Ibid., 151: entry for the 丁丑 day of the 5th month of Yongle 12 (23 May 1414); 152: entry for the 丁卯 day of the 6th month of Yongle 12 (12 July 1414).

92 Ibid., 158: entry for the 庚申 day of the 11th month of Yongle 12 (1 January 1415).
93 I only list a few of the many scholars who hold this opinion: Zheng Zhaojing 鄭肇經, "Discussions on Outlet Routes for Lake Tai (太湖出水路線的商榷)," *Water Conservancy* [中國水利] No. 2 (1982): 50–51; Wang Jian-ge, "From the Mouth of the Three Rivers to the Three Rivers," pp. 167–168.
94 Zhang Guowei, *Complete Record of Hydraulic Management in the Wu Region*, 14: 3.
95 Ibid.
96 Ibid., 4: 2.
97 Ibid., 15: 50.
98 Hai Rui 海瑞, *Precautions Against Forgetting* [備忘集], (Taipei: Shangwu yinshuguan, 1983), 1: 19–23.

Bibliography

Primary sources

Anonymous. *Jianwen Huangdi yiji* 建文皇帝遺蹟 [*Traces of the Jianwen Emperor*]. Beijing: Beijing daxue chubanshe, 1993.

Cao Shenfang 曹參芳. *Xunguo Zhengqi ji* 遜國正氣紀 [*Uprightness during the Abdication*]. Taipei: Mingwen shuju, 1991.

Chen Jian 陳建. *Huangming tongji jiyao* 皇明通紀集要 [*Essential Collection of the Comprehensive Imperial Records*]. Beijing: Beijing chubanshe, 1995.

Chen Jian. *Huangming tongji fachuan quanlu* 皇明通紀法傳全錄 [*Comprehensive Imperial Records and Laws*]. Beijing: Beijing chubanshe, 2005.

Chen Wei 陳威 and Gu Qing 顧清 (eds.). *(Zhengde) Songjiang fu zhi* (正德) 松江府志 [*The Gazetteer of Songjiang Prefecture (Zhengde Edition)*]. Taipei: Zhuangyan wenhua shiye youxian gongsi, 1996.

Feng Rubi 馮汝弼 and Deng Fu 鄧韍. *(Jiajing) Changshu zhi* (嘉靖) 常熟縣志 [*The Gazetteer of Changshu (Jiajing Edition)*]. Beijing: Shumu wenxian chubanshe, 1988.

Fu Yin 傅寅. *Yugong shuo duan* 禹貢說斷 [*Judgments about the Tribute of Yu]*). Taipei: Shangwu yinshuguan, 1983.

Gui Youguang 歸有光. *San Wu shuili lu* 三吳水利錄 [*Records of Water Conservancy in the Three Wu*]. Taipei: Shangwu yinshuguan, 1983.

Guo Songtao 郭嵩燾. *Shan Zhi shuwu wenji* 養知書屋文集 [*Collections from the Studio of Nourishing Knowledge*]. Shanghai: Shanghai guji chubanshe, 1996.

Hai Rui 海瑞. *Beiwang ji* 備忘集 [*Precautions Against Forgetting*]. Taipei: Shangwu yinshuguan, 1983.

Han Jun 韓浚. *(Wanli) Jiading xian zhi* (萬曆) 嘉定縣志 [*Gazetteer of Jiading County (Wanli Edition)*]. Taipei: Zhuangyan wenhua shiye youxian gongsi, 1996.

Huang Guangsheng 黃光昇. *Zhaodai dianze* 昭代典則 [*Rules of an Enlightened Reign*]. Taipei: Zhuangyan wenhua shiye youxian gongsi, 1996.

Huang Zuo 黃佐. *Gechu yishi* 革除遺事 [*Eradicating Things Past*]. Taipei: Mingwen shuju, 1991.

Jiao Hong 焦竑. *Guochao xian zheng lu* 國朝獻徵錄 [*Records of the Testimonies about the Worthies of the Reigning (Ming) Dynasty*]. Taipei: Mingwen shuju, 1991.

Lei Li 雷禮. *Guochao lieqing ji* 國朝列卿紀 [*The Great Officials of the Dynasty*]. Taipei: Mingwen shuju, 1991.

Lei Li. *Huangming dazheng ji* 皇明大政記 [*Imperial Politics*]. Taipei: Zhuangyan wenhua shiye youxian gongsi, 1996.

Li Jingde 黎靖德 (ed.). *Zhuzi yulei* 朱子語類 [*Master Zhu's Dialogues*]. Taipei: Huashi chubanshe, 1987.
Li Jinglong 李景隆. *Ming Taizhu shilu* 明太祖實錄 [*Veritable Records of Ming Taizu*]. Kyoto: Chûbun shuppansha, 1984.
Liu Sanwu 劉三吾 et al. (eds.). *Shu zhuan hui xuan* 書傳會選 [*Collected Selection of the Commentaries of the Book of Documents*]. Taipei: Shangwu yinshuguan, 1983.
Liu Tin-luan 劉廷鑾. *Jianwen xunguo zhi ji yuebiao* 建文遜國之際月表 [*A Monthly Chronology of the Abdication of the Jianwen Emperor*]. Taipei: Yiwen yinshuguan, 1970.
Mao Huang 毛晃. *Yugong zhinan* 禹貢指南 [*Guide to the Tribute of Yu*]. Taipei: Shangwu yinshuguan, 1983.
Mo Dan 莫旦. (*Hongzhi*) *Wujiangzhi* (弘治) 吳江志 [*The Gazetteer of Wujiang (Hongzhi Edition)*]. Taipei: Xuesheng shuju, 1987.
Qian Qianyi 錢謙益. *Guochu qunxiong shilüe* 國初群雄事略 [*Biographies of Major Figures at the Beginning of this Dynasty*]. Beijing: Beijing chubanshe, 1997.
Qian Sule 錢肅樂 and Zhang Cai 張采 (eds.). (*Chongzhen*) *Taicangzhou zhi* (崇禎)太倉州志 [*The Gazetteer of Taicang Subprefecture (Chongzhen Edition)*]. Edition of the 17th Year of the Kangxi Reign (1678–1679).
Qiu Jun 丘濬. *Daxue yanyi bu* 大學衍義補 [*Supplement to the Extended Meanings of the Great Learning*]. Taipei: Shangwu yinshuguan, 1983.
Sang Yu 桑瑜. *(Hongzhi) Changshu zhi* (弘治) 常熟志 [*Gazetteer of Changshu (Hongzhi Edition)*]. Taipei: Zhuangyan wenhua shiye youxian gongsi, 1996.
Shen Qi 沈棨. *Wujiang shui kao* 吳江水考 [*Investigation of the Waters in Wujiang County*]. Taipei: Zhuangyan wenhua shiye youxian gongsi, 1996.
Shen Shi-xing 申時行 et al. (eds.). *Minghui dian* 明會典 [*Laws and Institutions in the Ming*]. Beijing: Zhonghua shuju, 1989.
Shi Jian 史鑑. *Xicun ji* 西村集 [*Collections from Xicun*]. Taipei: Shangwu yinshuguan, 1983.
Tan Qian 談遷. *Guo que* 國榷 [*Discussions about the States*]. Shanghai: Shanghai guji chubanshe, 1996.
Tang Hezheng 唐鶴徵. *Huangming fushi bian* 皇明輔世編 [*Imperial Ministers*]. Taipei: Mingwen shuju, 1991.
Tu Shufang 屠叔方. *Jianwen chaoye huibian* 建文朝野彙編 [*Government and Public in the Jianwen Reign*]. Taipei: Zhuangyan wenhua shiye youxian gongsi, 1996.
Wang Ao 王鏊. *Gusu zhi* 姑蘇志 [*Gazetteer of Suzhou*]. Taipei: Shangwu yinshuguan, 1983.
Wu Zimu 吳自牧. *Mengliang lu* 夢粱錄 [*A Dream of Millet*]. Taipei: Shangwu yinshuguan, 1983.
Xia Chongwen 夏崇文. *Xia Zhongjing gong yishi* 夏忠靖公遺事 [*The Deeds of Xia Zhongjing*]. Taipei: Zhuangyan wenhua shiye youxian gongsi, 1996.
Xia Yuanji 夏原吉. *Zhongjing ji*忠靖集 [*Collected Works of Zhongjing*]. Taipei: Shangwu yinshuguan, 1983.
Xu Xueju 徐學聚. *Guochao dian hui* 國朝典彙 [*Collection of the Canons of Our Dynasty*]. Taipei: Zhuangyan wenhua shiye youxian gongsi, 1996.
Yang Shiqi 楊士奇 et al. (eds.). *Ming Taizong shilu* 明太宗實錄 [*Veritable Records of Ming Taizong*]. Kyoto: Chûbun shuppansha, 1984.
Yang Yi 楊儀. *Gu Wu wen lu* 古虞文錄 [*Literature from the Ancient Wu*]. Taipei: Zhuangyan wenhua shiye youxian gongsi, 1996.
Yao Wenhao 姚文灝. *Zhexi shuili shu* 浙西水利書 [*Documents of Water Management in Zhexi*]. Taipei: Shangwu yinshuguan, 1983.

Yin Shouheng 尹守衡. *Huangming shi qie* 皇明史竊 [*A Secret History of the Ming*]. Shanghai: Shanghai guji chubanshe, 1996.

Yong Rong 永瑢. *Qinding siku quanshu zongmu* 欽定四庫全書總目 [*Index to the Siku Quanshu*]. Taipei: Shangwu yinshuguan, 1983.

Zhang Guowei 張國維. *Wu zhong shuili quanshu* 吳中水利全書 [*Complete Record of Hydraulic Management in the Wu Region*]. Taipei: Shangwu yinshuguan, 1983.

Zhang Neiyun 張內蘊 and Zhou Dashao 周大韶. *San Wu shui kao* 三吳水考 [*The Waters of the Wu, Wuxing and Huiji Commanderies*]. Taipei: Shangwu yinshuguan, 1983.

Zhang Qin 張芹. *Bei yi lu* 備遺錄 [*A Record That Preserves What Might Be Forgotten*]. Taipei: Mingwen shuju, 1991.

Zhang Tingyu 張廷玉 et al. (eds.). *Mingshi* 明史 [*History of the Ming*]. Taipei: Zhonghua shuju, 1965.

Zhao Ying 趙瀛 and Zhao Wenhua 趙文華. *Jiaxing fu tuji* 嘉興府圖記 [*The Illustrated Records of Jiaxing Prefecture*]. Taipei: Zhuangyan wenhua shiye youxian gongsi, 1996.

Secondary literature

Bush, Simon. "Review of *Political Ecology: A Critical Introduction*, by Paul Robbins," *Geographical Research* 43: 2 (2005): 255–256.

Cai Taibin 蔡泰彬. "Mingdai wuchao yuanlao Xia Yuanji zhishui Jiangnan 明代五朝元老夏原吉治水江南 (The Elder Statesman of Five Dynasties, Xia Yuanji, and his Water Control in the Lower Yangzi Delta)," *Baisha renwen shehui xuebao* 白沙人文社會學報 [*Journal of Paisa Humanities and Social Sciences*] No. 1 (2002): 205–226.

Chan, Hok-Lam. "Legitimating Usurpation: Historical Revisions under the Ming Yongle Emperor (r. 1420–1424)," in Philip Yuen-sang Leung (ed.), *The Legitimation of New Orders: Case Studies in World History.* Hong Kong: The Chinese University Press, 2007, pp. 75–158.

Chao Zhongchen 晁中辰. *Ming Chengzu zhuan* 明成祖傳 [*Biography of Ming Chengzu*]. Taipei: Shangwu yinshuguan, 1996.

Ditmanson, Peter. "Venerating the Martyrs of the 1402 Usurpation: History and Memory in the Mid and Late Ming," *T'oung Pao* 93 (2007): 110–158.

Elman, Benjamin A. "The Formation of 'Dao Learning' as Imperial Ideology During the Early Ming Dynasty," in Theodore Huters, R. Bin Wong and Paulin Yu (eds.), *Culture and State in Chinese History: Conventions, Accommodations, and Critiques*. Stanford: Stanford University Press, 1997, pp. 58–82.

Fu Linxiang 傅林祥. "Wusongjiang xiayou yanbian xin jie 吳淞江下游演變新解 (A New Explanation of the Development of the Lower Reaches of the Wusong River)," *Xueshu yuekan* 學術月刊 [*Academic Monthly*] No. 8 (1990): 89–94.

Gao Shouxian 高壽仙. "Guanyu riben Ming Qing shehui jinjishi yanjiu de xueshu huigu—yi lilun moshi he wenti yishi shanbian wei zhongxin 關于日本明清社會經濟史研究的學術回顧—理論模式和問題意識嬗變為中心 (A Review on Japanese Research on Social and Economic History in the Ming and Qing—With a Focus on the Changes in Theoretical Models and Problems)," *Zhongguo jingji shi yanjiu* 中國經濟史研究 [*Research in the Chinese Economic History*] No. 1 (2002): 68–77.

Hamashima Atsutoshi 濱島敦俊. *Mindai konan nōson shakai no kenkyū* 明代江南農村社会の研究 [*A Study of the Rural Society in the Lower Yangzi Region in the Ming Dynasty*]. Tōkyō: Tōkyō daigaku shuppan kai, 1982.

Kawagoe Yasuhiro 川越泰博. *Mindai Kenbun chyô shi no kenkyû* 明代建文朝史の研究 [*History of the Ming Dynasty's Jianwen Reign*]. Tōkyō: Kyūko shoin, 1997.

Koyama Masaaki 小山正明. *Min Shin shakai keizaishi kenkyū* 明清社会経済史研究 [*A Study of Economic and Social History in the Ming and Qing*]. Tōkyō: Tōkyō daigaku shuppan kai, 1992.

Li Chaoran 李焯然 and Mao Peiqi 毛佩琦. *Ming Chengzu shilun* 明成祖史論 [*Ming Chengzu's Theory of History*]. Taipei: Wenjin chubanshe, 1994.

Liang Fangzhong 梁方仲. *Liang Fangzhong jingji shilun wenji* 梁方仲經濟史論文集 [*Collection of Liang Fangzhong's Works on the History of Economics*]. Beijing: Zhonghua shuju, 1989.

Lin Liyue 林麗月. *Mingdai de guozi jiansheng* 明代的國子監生 [*The Students of the Imperial Academy in the Ming Dynasty*]. Taipei: Sili Dongwu daxue Zhongguo xueshu zhuzuo jiangzhu wenyuanhui, 1978.

Man Zhimin 滿志敏. "Huangpujiang shuixi xingcheng yuanyin shuyao 黃浦江水系形成原因述要 (Causes of the Formation of the River Network of the Huangpu River)," *Fudan xuebao (shehui kexue ban)* 復旦學報 (社會科學版) [*Fudan Journal (Social Sciences Edition)*] No. 6 (1997): 54–59.

Mao Boke 茅伯科 and Tian Nanfan 田南帆. "Zheng He Xia Xiyang shiqi de Liujia Channel 鄭和下西洋時期的劉家港 (The Liujia Channel at the Time of Zheng He's Voyage)," *Shanghai daxue xuebao (shehui kexue ban)* 上海大學學報 (社會科學版) [*Journal of Shanghai University (Social Sciences Edition)*] No. 2 (1985): 71–76.

Okuzaki Hiroshi 奥崎裕司. *Chūgoku gōshin jinushi no kenkyū* 中国郷紳地主の研究 [*A Study of Chinese Gentry Landlords*]. Tōkyō: Kyūko shoin, 1978.

Robbins, Paul. *Political Ecology: A Critical Introduction.* Malden, MA: Blackwell Publishing, 2004.

Roth, Robin. "Review of *Political Ecology: A Critical Introduction*, by Paul Robbins," *The Canadian Geographer* 50.1 (2006): 135–136.

Wang Jiange 王建革. "Cong san jiang kou dao Sanjiang—Lou jiang yu Dong jiang de fuhui ji qi yingxiang 從三江口到三江—婁江與東江的附會及其影響 (From the Mouth of the Three Rivers to the Three Rivers—A Forced Interpretation of Lou and Dong River and its Influence)," in *Shehui kexue yanjiu* 社會科學研究 [*Social Science Research*] No. 5 (2007): 162–168.

Wang Jiange. "Song Yuan shiqi Taihu dongbu diqu de shui huanjing yu tangpu zhizha 宋元時期太湖東部地區的水環境與塘浦置閘 (The Hydraugraphic Environment and the Channel-Sluice System in the Lake Tai East Drainage Area in the Song and Yuan Period)," in *Shehui kexue* 社會科學 [*Social Sciences*] No. 1 (2008): 134–142.

Wang Jiange. "Shuiliu huanjing yu Wusong jiangliu yu de tian zhi (10–15 shiji) 水流環境與吳淞江流域的田制 (10–15世紀) (The Hydraugraphic Environment and the Field System in the Wusong Region (10th–15th Century)," *Zhongguo nongshi* 中國農史 [*Agricultural History of China*] No. 3 (2008): 3–15.

Wang Jiange. "Jing, bang fazhan yu Wusong jiang liuyu de weitian shuili (9–15 shiji) 涇、浜發展與吳淞江流域的圩田水利 (9–15 世紀) (The Development of Small Waterways and Water Management through Polder Dykes in the Drainage Area of the Wusong River [9th to 15th Century])," *Zhongguo lishi dili luncong* 中國歷史地理論叢 [*Collections of Essays on Chinese Historical Geography*] 24.2 (2009): 30–42.

Wang Jiange. "10–14 shiji Wuxong jiang diqu de hedao, weitian yu zhishui tizhi 10–14世紀吳淞江地區的河道、圩田與治水體制 (River Courses, Polders and Systems of Water Control in the Wusong Regions in the 10th–14th Centuries)," *Nankai xuebao (Zhexue shehui kexue ban)* 南開學報(哲學社會科學版) [*Nankai Journal (Philosophy, Literature and Social Science Edition)*] No. 4 (2010): 40–51.

Wang Ting 王頲. "Yuandai de Wusongjiang zhili ji gandao 'gailiu' wenti 元代的吳淞江

治理及幹道「改流」問題 (The Management of the Wusong River in the Yuan Dynasty, and the Problem of the Change of Course of the Main Waterway)," in *Zhongguo lishi dili luncong* 中國歷史地理論叢 [*Collections of Essays on Chinese Historical Geography*] 18.4 (2003): 83–91.

Wolford, Wendy. "Review of *Political Ecology: A Critical Introduction*, by Paul Robbins," *Annals of the Association of American Geographers* 95.3 (2005): 717–719.

Wu Yuming 吳聿明. "Zheng He xia Xiyang yu Taicang Yuan Ming chenchuang zhi yanjiu 鄭和下西洋與太倉元明沈船之研究 (Zheng He's Voyage and Shipwrecks in Taicang in the Yuan and Ming)," *Shanghai zaochuan* 上海造船 [*Shanghai Shipbuilding*] No. 2 (2005): 40–42.

Xie Shi 謝湜. "11 shiji Taihu diqu nongtian shuili geju de xingcheng 11 世紀太湖地區農田水利格局的形成 (The Formation of the Setup of Farmland and Water Conservancy in the Lake Tai Region, 11th Century)," *Zhongshan daxue xuebao (shehui kexue ban)* 中山大學學報·社會科學版 [*Journal of Sun Yatsen University. Social Science Edition*] No. 5 (2010): 94–106.

Xie Shi. "Taihu yidong de shuili, shuixue yu shehui (12–14 shiji) 太湖以東的水利、水學與社會 (12–14 世紀) (Water Conservancy, Hydraulics and Society East of the Taihu [12th–14th Centuries])," *Zhongguo lishi dili luncong* 中國歷史地理論叢 [*Collections of Essays on Chinese Historical Geography*] 26.1 (2011): 17–31.

Xie Shi. "Ming qianqi Jiangnan shuili geju de zhengti zhuanbian ji qi xiangguan wenti 明前期江南水利格局的整體轉變及其相關問題 (The Complete Transformation of the Setup of Water Conservancy in the Early Years of the Ming and Related Questions)," *Shixue jikan* 史學集刊 [*Collected Papers of Historical Science*] No. 4 (2011): 44–49, 72.

Yu Weiming 郁維明. *Mingdai Zhou Chen dui Jiangnan diqu jingji shehui de gaige* 明代周忱對江南地區經濟社會的改革 [*The Economic and Social Reforms in the Lower Yangzi Region Undertaken by Zhou Chen in the Ming Dynasty*]. Taipei: Shangwu yinshugan, 1990.

Zhang Xiaoping 張小平 and Zhou Shaoqing 周少青. "Zhengzhi hefaxing shiye zhong de Ming Chengzu—ge fanben de jiedu 政治合法性視野中的明成祖一個範本的解讀 (Ming Chengzu in the Field of Political Legitimacy—an Exemplary Reading)," *Jilin shifan daxue xuebao (renwen shehui kexue ban)* 吉林師范大學學報 (人文社會科學版) [*Jilin Normal University Journal (Humanities & Social Science Edition)*] No. 2 (2008): 59–64.

Zheng Zhaojing 鄭肇經. "Taihu chushu luxian de shangque 太湖出水路線的商榷 (Discussions on Outlet Routes for Lake Tai)," *Zhongguo shuili* 中國水利 [*Water Conservancy*] No. 2 (1982): 50–52.

Zhu Hong 朱鴻. *Ming Chengzu yu Yongle zhengzhi* 明成祖與永樂政治 [*Ming Chengzu and Yongle Politics*]. Taipei: Guoli Taiwan shifan daxue lishi yanjiusuo, 1988.

4 Media and the environment in treaty-port China

The "Woosung Bar" controversy in the 1870s

Marlon Zhu

Introduction

For forty years, beginning in the 1870s, the newspapers of Shanghai had reported on the silt deposits in the Huangpu River (黃浦江) near Wusong (吳淞, Woosung, as spelled in contemporary English sources) and the trouble it caused for access for steam shipping. The delays caused by the periodic buildup of silt became one of the most vexing problems for the Shanghai mercantile community, diplomats, and administrative authorities, both Western and Chinese. Earlier studies have missed the important role that English-language newspapers in Shanghai, mainly the *North China Herald* (hereafter *NCH*), played as one of the main public advocates encouraging dredging of these silt deposits on behalf of the merchant community. The newspapers highlighted the "Woosung Bar" question, making it a lasting public focus. Public opinion, enhanced by the press, tried to mobilize both consular officials and the Chinese Imperial Maritime Customs in hopes of maintaining Shanghai's leading status as the largest emporium in the China trade. This chapter explores how the "Woosung Bar" crisis was perceived, prioritized, and remedied in order to reveal the agency of the commercial press, which represented the cultural politics of the "shipping-insurance hegemony" in nineteenth-century coastal China.

The role of the media in representing environmental issues has never been more evident than in the climate change debates of recent years. A systematic study by Maxwell T. Boykoff analyzed how the media filtered the voices of people raising awareness on the issue of climate change. He argued that the media served "a vital role in communication process between science, policy and the public" and that "[the media] has stitched together climate science, governance and daily life."[1] Boykoff examined various forms of media, including newspapers, books, television, film, radio, and the Internet, in his study. This chapter takes a similar approach by considering a hydrographical issue facing the mercantile community in nineteenth-century China—the silt buildup in the Huangpu River and the occasional blockage of the only shipping channel to Shanghai—through the framing of the issue in Shanghai's English-language newspapers. In the process, the media in nineteenth-century China created a discursive space between the often antagonistic mercantile and bureaucratic agencies.

Scholars have recently become interested in the role of media in modern China. They discovered an important role for media since newspapers began in China in the nineteenth century. In the 1830s, the earliest English-language newspapers, established in Canton, were responsible for fermenting bellicose sentiment among foreign (mainly English) traders, advocating a war against the Chinese government and contributing to the outbreak of the Opium War (1839–1842).[2] Nonetheless, most studies on media in modern China put their emphasis on the burgeoning Chinese-language newspapers published since the 1870s. Major arguments focus on the media's role in cultivating a Chinese "public sphere" and its significance to Chinese nationalism or a culture of mass consumption.[3] Almost none of those studies—including those produced in China—deal with the media's coverage of the environment.[4] In the 1870s, the newspapers' persistent coverage of the "Woosung Bar" silt deposits kept an environmental issue alive as one of the important concerns within the mercantile community, which principally relied on the shipping channel. The discursive process surrounding the "Woosung Bar" issue—as well as the economic, political, and social context of treaty-port China—suggests an important relationship between the media and the environment.

A "heavenly sent barrier": the origins of a propaganda campaign

The "Woosung Bar" was one of the most important hydrological issues in the history of Shanghai.[5] The underwater sand bar greatly threatened the status of the city as the central entrepôt of the East Asian maritime economy. Since opening to foreign trade in 1843, Shanghai soon grew to be the largest among the treaty ports. Geographical setting helps account for the rise of Shanghai in the age of steam ships. Shanghai was situated in the crossroad of China's coast line and the steam-navigable Yangtze River. However, Shanghai does not lie exactly in the estuary of the Yangtze. The Huangpu River connects the Yangtze with Shanghai.

Like every river in the world, not every part of the Huangpu was safe for steam vessels. One of the major silting sites in Huangpu was located near Woosung, a town at Huangpu's estuary. In the nineteenth century, Gough Island (高橋沙, see Figure 4.1) near Woosung had divided the Huangpu River into two channels. The wider channel on the left bank, the "Junk Channel," was so shallow that only Chinese junks could navigate it. The right bank channel was called the "Ship Channel." Although narrower than the Junk Channel, the Ship Channel was deeper and foreign vessels with a greater draft could pass. Despite Gough Island dividing the river, the Huangpu had other sand buildup problems; the most agitating to the shipping community was the Woosung bar, which appeared at the entrance of the Ship Channel. Its proximity to Woosung gave the silt sediment its name, "The Woosung (Inner) Bar."

The changing nature of the bar might be the reason for the prolonged four-decade controversy, which lasted from the 1870s until 1910. In a sailing guide

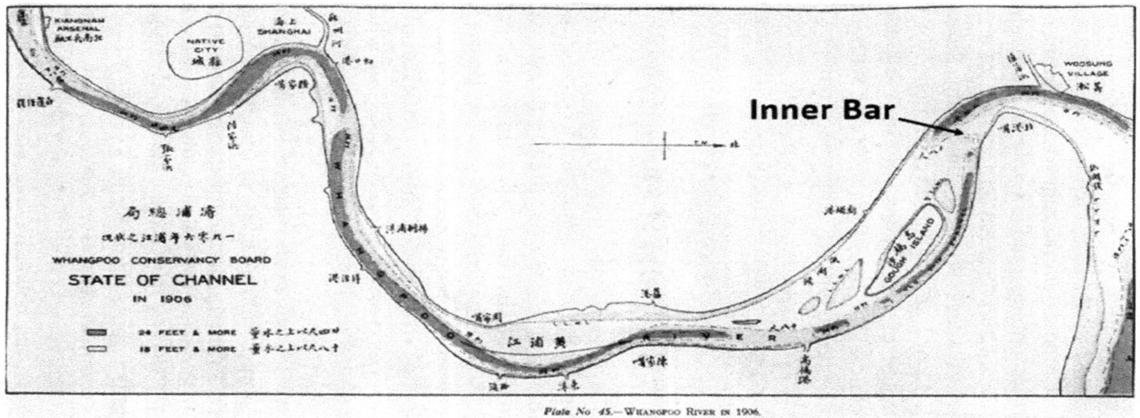

Figure 4.1 The channel of the Huangpu River in 1906 and the bar (source: Whangpoo Conservancy Board, *The Port of Shanghai. General Series. Report No. 8*. Shanghai: Whangpoo Conservancy Board, 1929).

Note
North is at the right. The darker area denotes where the river depth was over twenty-four feet.

published by the British Admiralty in 1863, ship captains and commanders were warned that "it would be imprudent for a stranger to enter the Wusong [Woosung] river without a pilot, who is always in attendance at the entrance, for banks within are constantly undergoing changes from the alluvial deposits."[6] The constant shifting of silt deposits was not visible from the banks but took place unseen under the water. In the last decades of the century, people finally discovered the cause for the formation and seasonal changes in the Woosung Bar. They found that it was caused by a great quantity of silt being brought in by flood tides from the Yangtze during the spring and summer months (May to September), but was reduced at other times of the year with considerable regularity.[7]

The existence of the Woosung Bar was noticed by one of the earliest foreign visitors. Under the order to conduct a coastal survey assigned by the British East India Company, H. H. Lindsay (1801–1881) and Karl Gützlaff (1803–1851) observed in June 1832 that they were confident that their ship, the *Lord Amherst*, could safely pass the bar, which "presents no dangers whatever, as will be apparent from Captain Rees's chart, and the guides for entering are extremely simple." They also recorded that at its lowest point water was always found to be four fathoms (24 feet) over the bar, concluding that the Woosung [Huangpu] River was "in every respect one of the finest and most navigable in China."[8]

These earlier foreign passengers' hydrographical intelligence of the bar was accurate for the time. In the 1830s, the bar's seasonal shifting depth seemed never to threaten vessels of any size. This continued to the case over the following two decades (1840s and 1850s), while the average tonnage of vessels frequenting Shanghai was 1,000 tons with an average draft of 12–13 feet, far from the bottom of 24 feet. In later years, the bar gradually became a foreseeable problem as the size and complement of ships continued to increase in the 1860s, when the average ship size grew to 2,300 tons with the average draft 16–17 feet.[9]

During the 1860s, when the bar was not troublesome to vessels, foreign shipping firms had already expressed their concerns to the Chinese government of the need for a dredging. The *NCH* reported in 1901 that that the foreign mercantile marine started petitioning Robert Hart (1835–1911) about the dredging as early as 1863. Hart was then at Shanghai and had just been promoted to be Inspector-General (that was, the head) of the Chinese Imperial Maritime Customs. In response, one of Hart's subordinates, the Harbor Master of Shanghai (J. M. Hockly) recommended that the merchants should raise the necessary funds to deepen and conserve the channel.[10] Meanwhile, the Customs continued to take some measures, short of dredging, to facilitate navigation over the bar. For instance, beacons and buoys were established under the auspice of the Harbor Master to mark the exact position of the bar. Moreover, a tide pole and signals hoisted on a flagstaff were established to indicate the timely depth of the water on the bar.[11]

Despite these navigational aids provided by the Customs, the Woosung Bar remained fuel for tensions between the Shanghai mercantile community and Hart, after he moved the Inspector-General's office from Shanghai to Beijing in

1865.[12] The constant bar-monitoring and other efforts by the Customs were far from satisfying to the merchants whose steamers frequented Shanghai. After the Taiping Rebellion (1850–1864), tonnage dues collected by the Customs became an imminent issue discussed in the local newspapers. The mercantile community reiterated their demand to dredge the Huangpu River, using the funds from the tonnage duties, which were formerly appropriated by the Chinese government to fund the war effort against the Taipings.[13] In 1868, responding to such remonstrations, Hart established a Marine Department within the Customs. The new department was to facilitate coasting and inner-river navigation in China using funds from tonnage duties. However, to the disappointment of the Shanghai mercantile community, the demand to dredge the Woosung Bar seemed to be edged out by the Department's ongoing wider scheme for establishing lighthouses and lightships along the China coast.[14]

When commenting on Hart and the Marine Department's ambitious lighthouse project, the mercantile community in Shanghai continued to express, through the press, the necessity of dredging. Their demand repeatedly occupied the columns of the local newspapers for decades (see Figure 4.2). The Shanghai merchants believed that dredging the bar was a legitimate use of tonnage dues. They also argued that Article 32 of the Treaty of Tientsin (1858), which was signed by the Chinese government and several Western powers, stated that "the Consuls and Superintendents of Customs shall consult together regarding the erection of beacons of lighthouses and the distribution of buoys and lightships,

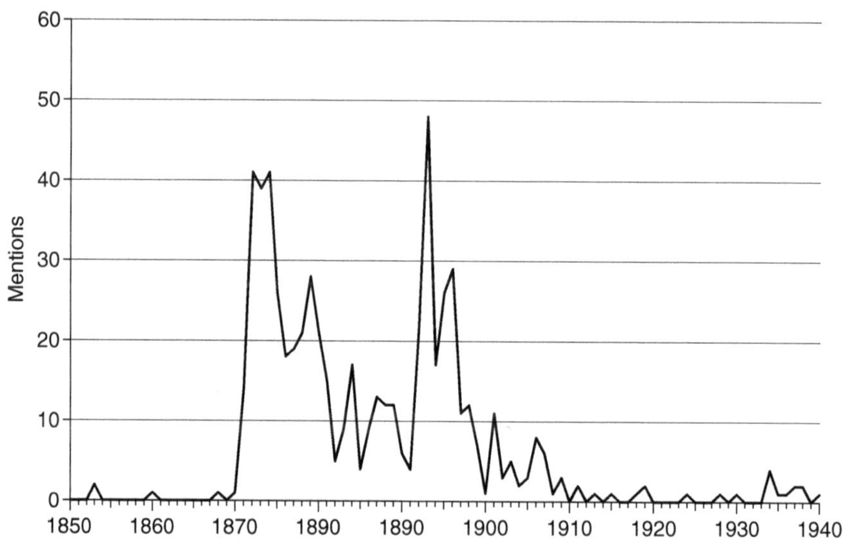

Figure 4.2 Mentions of "Woosung Bar" in the *North China Herald*, 1850–1940.

Note

From the *North China Herald* database made by Brill & Co., which lacks specific years (1857–1859, 1866, 1867, and 1925). The number of original mentions of "Woosung Bar" is 625. This figure only counts 612, which excludes references to the bar in the *Herald's* yearly indices.

as occasion may demand."[15] The dredging, the Shanghai press and merchants opined, fell under this article of the treaty as work for assisting navigation. This legal reasoning underpinned many of the petitions sent to Chinese and foreign authorities.

In addition to offering a broader interpretation of the treaty, the Shanghai press had also launched a propaganda campaign which attached the need of dredging to then prevailing discourse of progress. When there was no official response to the mercantile community's demand for dredging, the press started to stigmatize the Chinese authorities as standing still at a critical historical moment. Newspapers implied that since the Woosung Bar did not interfere with the navigation of Chinese junks, the Chinese government had no difficulty seeing it as none of their business. According to the Shanghai press, the Chinese authorities in Beijing considered the bar to be a "heavenly sent barrier," a natural gift to be kept rather than removed. It could prevent foreign ironclads or any other large men-of-war from entering Shanghai, an important security concern of the Chinese that prevented any official moves in support of dredging. To respond to these concerns, the Shanghai merchants announced that the dredging was a criterion for China to be entered onto the lists of "civilized countries." Dredging the Woosung Bar was "internationally" important both because of China's treaty obligations and as a responsibility of a civilized country. The mercantile remonstrations had even reached their home governments in Europe and the United States. In 1874, Hart opined that the situation had become even worse. While the Chinese government had been having difficulties with Japan over Formosa, the Woosung Bar question had nearly become another *casus belli* against China by other foreign powers.[16]

Initiatives of the "shipping-insurance hegemony" and the press

The Shanghai newspapers reported meticulously on this enduring forty-year agitation between the merchants and the authorities. This long spotlight had reflected the mercantile politics in the treaty-port Shanghai. The term "Woosung Bar" appeared a total of 612 times in the *NCH* between 1870 and 1910 (see Figure 4.2). Within this period there were two peaks of interest in the Woosung Bar, the 1870s and the 1890s. This chapter will focus on the 1870s, for it not only represented the origin of the Woosung Bar controversies, but also demonstrates the press's initiative in establishing the discourse over the bar.

The first explosion in interest in Shanghai newspapers over the Woosung Bar appeared between 1872 and 1874 (see Figure 4.2) and was dotted by detailed reports of mercantile initiative and early insistence on the necessity of dredging. Earlier efforts to encourage dredging in the 1860s do not appear in the newspapers, suggesting that the bar was not causing too many troubles for merchants. All the controversies over the bar commenced in 1870. The single appearance of the "Woosung Bar" in the newspapers in 1870 was an editorial in the *NCH*. While commenting on the lighthouse works done by the Customs' Marine

Department, the editorial suggested that further actions on the bar should be taken by the local mercantile community. It advocated that the Shanghai General Chamber of Commerce should extend their previous efforts upon the bar, since "much local good may be accomplished by local pressure." According to the *NCH*, at least three sections of the community would benefit from removing the bar. The most obvious beneficiary would be the steam shipping companies. Second, the marine insurance offices, which took risk on the cargoes and were chiefly owned by Shanghai shareholders, would see a reduction in their liability if the Huangpu could be navigated more safely. Third, the general Shanghai commercial firms would benefit because a considerable portion of their cargo crossed over the bar.[17]

The Shanghai press's coverage on the Woosung Bar had revealed the dominance on cultural representation of this combined interest of shipping, insurance, and general commercial firms. Analyzing the language used by the newspapers to describe the Woosung Bar controversy is to examine a series of common actions by the dominant merchants in treaty port China. I call this group, as well as the activities they engaged in, the "shipping-insurance hegemony."[18] These merchants had been the major readers and supporters of the commercial press through purchasing advertisements and subscriptions. The allegiance between merchants and media began from the first day of the English-language newspapers in China. James Matheson (1796–1878), one of the co-founders of the renowned Jardine, Matheson & Co., published the *Canton Register* in Canton from 1827. Prices of then illegally imported opium were bulletined publicly in the papers. In the case of the Woosung, the *NCH* continued to serve commercial and navigational interests by manufacturing the image of impending peril, and crying out for the necessity of official operations, including dredging.

The 1860s was the watershed for this specific capitalist enterprise for it witnessed the booming of China trade and the mushrooming of foreign steam navigation and marine insurance companies in China. Existing studies of nineteenth-century commercial history of China have focused mainly on the business of steam shipping. Few, however, touched on their usually intimate relationship with the business of marine insurance. Instead of manufacturing or other business such as mining and railroad building, steam shipping constituted the key sector in China's "mercantile capitalism" in the nineteenth century.[19] Several factors explain the new opportunity and situation. International treaties signed after the Second Opium War (1856–1860) had opened more ports along the Yangtze and the northern ports of China (such as Tianjin) for foreign merchants. The London *Times* reported in 1864 that the import trade of Shanghai had increased nearly twofold in just the three years between 1860 and 1863.[20] Days later, the same London newspaper commented that "the present El Dorado of commercial men seems to be China."[21] The "China rush" in the 1860s was reaffirmed by a British Consul in Shanghai, who reported in 1867 that the number of foreign firms in the city increased threefold in the previous few years.[22] At the end of the 1860s, the growing trade was also enhanced by the improvement of communication such as the submarine telegraph around 1870 and the opening of the Suez Canal in 1869. No wonder scholars had claimed that in 1860s there was a commercial revolution in China.[23] Meanwhile, the 1860s rush had increased the importance of Shanghai as the greatest emporium in the China trade. In a long-term perspective, Hosea Ballou Morse observed that in 1874, 60 percent of the China trade flowed through Shanghai. This amount only fell to 55 percent by 1902.[24]

The booming China trade in the 1860s explained the timing of agitation of the shipping-insurance hegemony over the bar in the early 1870s. In November 1870, the Shanghai General Chamber of Commerce sent, through Foreign Consuls in Shanghai, a memorial to the Chinese magistrate of Shanghai (Shanghai Daotai 道台) on the condition of the bar. They suggested the Chinese should begin an ordinary dredging in order to make the bar navigable to vessels with a draft of twenty-four feet and thus to prevent the almost daily occurrence of costly delays and accidents.[25] In February 1871, the Harbor Master of Shanghai and the Division Inspector of the Customs, S. A. Viguier, surveyed the bar in response to these demands. He found a new channel, 500 feet wide, cutting through the Woosung Bar, which may have been caused by the narrowing of the Ship Channel. Its depth was no less than fifteen feet during the lowest tides of the year. Adding that the rise and fall of the tide was never less than six feet, Viguier concluded that vessels drawing under twenty feet would always be able to navigate the bar. Since the mean rise of the tide was ten feet, it would only be under exceptional circumstances that vessels drawing 20–23 feet would be prevented access by the bar. The foreign consuls in Shanghai invited Danish, British, French, and U.S. commanders of vessels-of-war to verify Viguier's survey, concluding that this new channel would apparently obviate the necessity of dredging.[26]

Unfortunately, due to the shifting nature of the bar, Viguier's happy finding did not close the controversy between the mercantile community and the Customs' hydrographical officers in Shanghai. In addition, at the end of 1869 more large vessels frequented Shanghai due to the opening of the Suez Canal. The canal had caused the diminishing of the transshipments that formerly took place, and allowed larger vessels to sail to the Far East directly. This may explain the timing of two further, nearly identical, petitions regarding the bar originating from the shipping-insurance hegemony.

These two petitions came from the two largest mail-steamer companies in Shanghai, the British Peninsular & Oriental Steam Navigation Company (P & O) and the French Compagnie des Services Messageries Maritime.[27] In September 1871, S. J. G. Jellicoe, a P & O agent in Shanghai, wrote a letter to the Shanghai General Chamber of Commerce. He asked the Chamber to pay attention to "anomalies" in the Woosung Bar, which had recently detained a P & O steamer (the *Emeu*) at Woosung, causing a twelve-hour delay. The Chamber then forwarded the letter to the Consular Body of treaty powers in Shanghai.

Jellicoe accused Viguier's survey, although internationally confirmed, of being rife with falsehoods. The P & O's own survey revealed that the depth of the water on the Bar rarely exceeded 17–21 feet, lower than Viguier's estimates of 22–28 feet. The previous May, Jellicoe wrote Viguier requesting data on water depths. Viguier held that the depth of the water had not changed since his February survey. He added that in May

> the New Channel over the Woosung Bar is 14 feet 6 inches, at dead low water spring tides. The rise and fall (influenced by the wind) is from 12 to 14 feet in springs and from 8 to 10 during neap tide, therefore the depth of water on the Bar ranged from 22 feet to 28 feet.[28]

Receiving the confirmation from the Harbor Master, Jellicoe accordingly informed the P & O Directors that the Company's larger mail-steamers could then run direct to Shanghai, since the Woosung Bar, as a contingent risk before, had been removed by nature itself. However, the P & O Steamer *Emeu*, which Jellicoe claimed was drafting only eighteen feet at that time, was still detained by the bar on September 9.

For shipping companies, even a temporary detention of large vessels meant huge costs even if there was no accident or physical damage. In addition to valuable cargo being delayed, the vessels themselves had heavy daily operating expenses. Take the *Emeu* for instance, this home-going ocean steamer then carried not only "homeward" mail, but a large and valuable cargo, worth around 1,600,000 taels. In the month following the *Emeu* case, there were two incidents of P & O steamers being stopped by the Woosung Bar (Table 4.1).[29]

It seemed justifiable for Jellicoe to petition the Chamber of Commerce, urging them to take steps to improve navigation over the bar, either by dredging or other means. He continued to warn that

> as it is manifest that if this apathy on the part of the Chinese authorities [the Harbor master and the Customs implied] is allowed to continued, it will soon be found to have a serious effect on the shipping and trade of this Port.[30]

Ch. De Crety, the agent of another state-chartered mail-steamer company, the French Messageries Maritimes, saw the situation a bit differently from his P & O colleague. However, the French agent furnished a recent survey made by a French captain (Captain Varagot of the Messageries' mail-steamer *Phase*), in which the accuracy of the Customs' signal of the depth was greatly challenged.[31]

The two steam-shipping giants' accusations soon got a reply from both the foreign and Chinese authorities. The Chamber of Commerce forwarded Jellicoe's petition to the Consular Body of foreign powers in Shanghai. The Chamber asked the consuls to once again apply to the Chinese authorities, for it might result in the removal of "this source of anxiety [the Bar]."[32] In a meeting of treaty consuls of Shanghai held on October 16, 1871, George F. Seward, the Senior Consul of the Consular Body and the U.S. Consul General in Shanghai, presented the petitions from the P & O and the Messageries Maritimes. Seward mentioned that he had discussed the issue with the Customs Commissioner of

Table 4.1 P & O steamers detained by the Woosung Inner Bar, September 1871 to January 1872

Year	*Month*	*Date*	*Steamer*	*Description*
1871	September	9	*Emeu*	With outward mails, detained 12 hours; draft 19 feet
	September	11	*Behar*	With inward mails, detained 20 hours; draft 18 ft. 8 in.; depth of water signaled on Bar 18 ft. 6 in., but found only 17 ft. 9 in. in mid-channel.
	September	23	*Behar*	With outward mails, detained 12 hours, and damaged her rudder from scraping over the Bar, and had to dock in Hongkong in consequence; draft 18 ft. 3 in.; 17 ft. signaled.
	October	26	*Australia*	Lost a freight of Taels 8,000 to Taels 10,000 for Rice to Whampoo [Huangpu] in consequence of not being able to cross the bar, drawing 22 ft. 6 in.
	November	7	*Sunda*	With inward mails, detained 12 hours, and had to be lightened at Woosung; draft 18 ft. 6 in.
1872	January	5	*Mirzapore*	Lost local freight of about Tls. 3,000 (her draft would be about 20 ft.) and had to leave the anchorage on the afternoon of the 4th in order to cross the Bar, which entailed the hiring of a Tug to take the mails &c. to Woosung, and other expenses.
	January	22	*Sunda*	With a full cargo, and drawing 19 ft. 6 in., was detained in port from 22 to 24 January
	January	22	*Peking*	Drawing 18 ft. 9 in., with inward mails, was delayed 24 hours.
	January	25	*Malacca*	Drawing 21 ft. 4 in., lost two days, and was obliged to return to Shanghai to discharge part of her cargo.

Several other petty cases of delay and inconvenience have arisen, and have resulted in loss to the Company.

Source: Dispatch of Seward (U.S. Consul General in Shanghai) to the U.S. (Acting) Secretary of State, No. 521, February 21, 1872, Enclosure No. 7, reported by the P & O.

Shanghai, Thomas Dick, and Viguier, the Harbor Master. They replied to the mail-shipping companies' petitions and stated the facts of the case to the U.S. Consul. The two Customs officers held that it appeared that the water on the bar was then the same as that shown by the Admiralty Charts from the 1862 survey (eleven and twelve feet at low water spring tides), but unfortunately, the greater depth (fourteen feet six inches) reported in February 1871 no longer existed.

The seasonal nature of the Woosung Bar's silting was acknowledged by the Customs officers. "[T]he shoaling commenced probably in June, and has progressed until the present moment [October]. Viguier added that, the silt may cut away, after the season of high water in the Yangtze has passed." And Viguier was confident that his February survey was correct at that time, since a naval committee appointed by the Consuls had verified his survey then. In reply to the challenge of the accuracy of the Customs' signal of the Bar, Viguier stated that great care had been taken to make the signals at the Woosung Station correspond to the actual depth of water, as established by sounding day to day. Moreover, Seward added that he had spoken about the bar to Robert Hart, the Inspector-General of the Customs. Seward was then informed by Hart that "a dredging machine would be sent for."[33]

The official replies from both the treaty consuls and the Customs officers did not satisfy the Shanghai mercantile community. No substantial steps were taken to remedy the anxiety while more cases of detained ships occurred during the winter of 1871–1872 (see Table 4.1). A second remonstration came the same winter from seventy-seven commercial firms in Shanghai. This petition was initiated by Herbert S. Morris, the Secretary of the North-China Insurance Company.[34] This petition was a reflection of the shipping-insurance hegemony, co-signed by other major insurance offices, shipping companies, banks, and commercial firms of Shanghai. Although the full list of signatories was not carried in the press, it was provided in a dispatch of the U.S. Consul in Shanghai to Washington, D.C. (Table 4.2). In the petition, the mercantile community complained that there were no practical results since they first brought the subject before the treaty consuls and the Shanghai General Chamber of Commerce. They urged the Customs Commissioner of Shanghai, Thomas Dick, of "the importance and necessity for immediate steps being taken to remedy the insufficient water on the Bar at certain seasons."[35]

The media enhanced local pressures, pushing them far beyond Shanghai. Almost at the same time the merchants' petitions went to Beijing and London. In February 1872, the Shanghai merchants' petition was sent to the Chinese government and Foreign Ministers in Beijing, acknowledging the urgency of the obstruction caused by the bar. The message arrived at Beijing by two channels. Herbert S. Morris asked Thomas Dick, the Customs Commissioner of Shanghai, to forward the letter to his supervisor in the Customs, Inspector-General Robert Hart. Meanwhile, the Shanghai Chamber of Commerce addressed their version, which enclosed Morris' January petition, via the U.S. Consul General in Shanghai, to the Foreign Ministers in Beijing. In these remonstrations to Beijing, the Shanghai merchants complained that earlier petitions to the Chinese authorities (respectively in the autumn of 1870 and February 1871) had proved to have been in vain. The overlooked Shanghai merchants warned that if the apathy of the Chinese government continued, it would lead to "serious permanent injury of the port as the central entrepôt for the Northern Provinces of China." And since "the cost of maintaining an efficient channel by means of dredging would be so small, as compared with the amount of dues collected from shipping," it was reasonable for tonnage dues to be devoted to the maintenance and improvement of navigation.[36]

The press played a role here by exaggerating the seriousness of the situation at the Woosung Bar. The same petition from the Shanghai Chamber to Beijing, which was not carried publicly in the *NCH*, admitted that the mail steamers of the Messageries Maritimes had not been prevented from passing the bar, as their departure schedule had been arranged at periods of highest tides. In addition, it seemed that the water level above the bar was not deteriorating. At the very least it was not worsening so quickly to become an immediate danger to navigation in the 1870s, as argued by agents of P & O, the Secretary of the North-China Insurance Company, and the Shanghai press.

"Apologists for the Chinese stagnation" in Beijing

With such uncertainty about the changing depth of the water around the Woosung Bar, it was not surprising that the official reply from Beijing continued to frustrate the Shanghai mercantile community. Fredrick F. Low, the U.S. Minister in Beijing, told the Shanghai merchant community that Robert Hart "appears to be doing all he can, with the means at his command, to meet the wishes of the mercantile community at Shanghai concerning the Woosung Bar. A dredging machine has been ordered in England for that work." In addition, the U.S. Minister regretted that

> a scientific and thorough examination, by person thoroughly competent for such work, had not made of the Bar in question before proceeding to expend the funds which the maritime interest can ill afford to see spent upon works of doubtful utility.

In contrast to Shanghai's lack of "scientific" information regarding the bar, the U.S. Minister informed Shanghai that Hart had submitted to him the reports of the examinations Hart had ordered, as well as the plans and estimates for the removal of the bar. Even though, the U.S. Minister still doubted Hart's scheme due to a lack of sufficient data required to undertake such a project. Based on the Minister's personal experience on similar cases in the United States, Low held that it would be very difficult, or rather impractical, to permanently deepen the channels of such rivers simply through dredging.[37]

Frederick Low was as well informed on hydrological questions as he was on the source and composition of the voices lobbying Hart. They had demonstrated the influence of the shipping-insurance hegemony exerted from Shanghai. The Minister noted:

> Were I permitted to make a suggestion in regard to this subject, I should say those most largely interested in commercial pursuits, in conjunction with the owners and agents of the various transportation lines which centre at Shanghai should, in their own interest, and also in the public interest, cause to be made such an examination such as I have suggested. If the work is impracticable, the cost of experiments will be saved, and ship-owners and ship builders will know the actual facts; if practicable, then there would be some *reliable data* on which to base representations which may be necessary to make to the Government in Peking [Beijing].

Table 4.2 Signatories of the petition of January 11, 1872

	Company or individual	*Agent or secretary for*
I Marine Insurance Offices and Shipping Companies		
1	Herbert S. Morris	North-China Insurance Co.
2	Jardine, Matheson & Co.	Canton and other Insurance Offices
3	Gibb, Livingston & Co.	Lloyds
4	Augustine, Heard & Co.	China Traders and other Marine Insurance Co.'s
5	Siemssen & Co.	Œsterling Sea and Fire Insurance Co. and Samarang do.
6	F. D. Barnes	The P. & O. Steam Navigation Co.
7	A. Hennequin (Acting Agent)	Messageries Maritimes Co.
8	Geo. F. Bowman	Pacific Mail Steam Shipping Co.
9	Butterfield & Swire	British and Foreign Marine Insurance Co. and The Ocean S. S. Co.
10	Samuel Brown	Union Insurance Society of Canton
11	Olyphant & Co.	China Insurance Co. Merchants' Mutual Marine Insurance Co. of San Francisco Board of Underwriters Union Steam Navigation Co. and Comagnie Russe de Nav. & c.
12	Russell & Co.	Yangtsze Insurance Association
13	Gilman & Co.	Various London Marine Insurance Companies
14	W. S. Schmïdt	China and Japan Marine Insurance Co.
15	Chapman, King & Co.	Hongkong Insurance Co.
16	Textor & Co.	Home Colonial Marine Insurance Co.
17	Smith Archer & Co.	The Borneo Co.
18	Holliday, Wise & Co.	London Assurance Corporation
19	Frazar & Co.	New York and Boston Underwriters American Shipmaster's Association National Marine Insurance Co. Australia Java Sea and Fire Insurance of Batavia
20	Wm. Pustau & Co.	Hamburg and Bremen Underwriters
21	Shaw Brothers & Co.	Union Marine Insurance Co. of Liverpool and London
22	Scheibler & Matthaei & Co.	Swiss Lloyds

Individual	Agent or manager for bank	
II Banks		
1 Robert Fergusson	Chartered Mercantile Bank of India, London and China	
2 D. MacLean	Hongkong & Shanghai Banking Corporation	
3 A. Paterson	Oriental Bank Corporation	
4 A. Kauffmann	Competoir d'Escomte de Paris	
5 L. W. Mullins	Agra Bank, Limited	
6 J. G. Marshall	Chartered Bank of India, Australia & China	
III Other Commercial Firms		
1 Turner & Co.	18 p.p. F.R. Gamwell, H. R. Hearn	35 W. Reme & Co.
2 Alfred Dent & Co.	19 Carter & Co.	36 J. Brdadhusrst Tootal
3. Brand Brothers & Co.	20 Rothwell, Love & Co.	37 Findlay, Wade & Co.
4 Thorne Brothers & Co.	21 MacLean, Wallace & Co.	38 Hogg Bros.
5 Little & Co.	22 F. W. Muller & Co.	39 Reiss & Co.
6 Bower. Hanbury & Co.	23 Overbeck & Co.	40 Bovet Bros. & Co.
7 Westall, Brand & Co.	24 Bourjau, Hubener & Co.	41 C. J. Skeggs &Co.
8 Blain & Co.	25 Rodewald, Schonfeld & Co.	42 Dickinson & Co.
9 Tate & Hawes	26 Milson & Tod	43 J. Jarvie & Co.
10 Dent & Co.	27 Bull, Purdon & Co.	44 J. C. Helbling & Co.
11 Reid & Co.	28 Wright, Burkill & Co.	45 Framjee Hormusjee & Co.
12 A. Provand & Co.	29 Adamson, Bell & Co.	46 M. D. Ghandy & Co.
13 Lindsay & Head	30 E. Schellhass & Co.	47 Primrose & Co.
14 Weston & Co.	31 Telge, Nolting & Co.	48 p.p. Cawasjee Pallanjee & Co., J. Rustomjee
15 Birley, Worthington & Co.	32 Morris, Lewiss & Co.	49 A. Droege & Co.
16 Coutts & Co.	33 G. A. Monro & Co.	
17 David Sasson, Sons, & Co.	34 Geo. Barnett & Co.	

Source: Seward (U.S. Consul General in Shanghai) to the U.S. (Acting) Secretary of State, No. 521, February 21, 1872

The absent of "actual fact" and "reliable data" implied that the narrative of deterioration of the Shanghai press was questioned by the authorities in Beijing and became their official excuse for not dredging. Nonetheless, the local press criticized the U.S. Minister's suggestion that the Shanghai merchants fund their own survey. The *NCH* accused Low of taking up a position of an "apologist for Mr. Hart." "Foreign merchants do not hold the Inspector-General of Customs responsible, but the Government." Newspapers were particularly opposed to Low's suggestion that the survey come at their own expense, since they already paid the tonnage due which was more than enough to fund the study.[38]

The British Minister in Beijing, in the eye of the press, seemed to hold the same "apologist" position as his U.S. counterpart. Thomas France Wade wrote a letter to the Shanghai merchants about the bar, acknowledging that a proposal to provide the proper machinery required had been for some time before the Zongli Yamen (總理衙門, Chinese Foreign Office) and that Hart had been instructed to give early attention to the matter. This, however, was not news to the Shanghai merchants, who had known of these developments five months earlier.[39] Moreover, according to the newspapers, Wade seemed to be a spokesperson for Customs, merely repeating that the tonnage dues had already been spent to facilitate navigation through the project to illuminate and buoy the coast of China.[40]

Both the U.S. and British Ministers' replies were heavily criticized by the Shanghai press as being parrots for Robert Hart. Continuing to insist on the immediate necessity of dredging, the *NCH* criticized those diplomats in Beijing as shifting from being "advocates of progress" to apologists for "Chinese stagnation." "There must surely be something in the atmosphere of Peking, which at times makes Foreign Ministers appear to act as drags on the progressive tendencies of their nationals."[41] One of the factors that contributed to such a retrograde atmosphere, the *NCH* continued, was Hart and his foreign Customs, pointing out that the Customs had become a buffer between the Chinese government and Shanghai merchants. "If the Chinese Government elect to work through their [the Customs'] agency, well and good; but if it fail to act at all, it is not the Customs but the Chinese Government we blame." The press then defended their position, arguing that the situation in Woosung was becoming a greater daily nuisance due to the increasing number of steam vessels frequenting Shanghai.

In the same issue responding to Beijing, the newspapers confessed that the bar itself was not deteriorating, but merely subject to the normal seasonal changes. To the claims that the dredging was not practical, the newspapers argued that it was a "common sense conclusion" that the capability of a dredger to move silt compared to the rate of accumulation on the bar was "an equation capable of solution, so equally can the horse-power required to remove it be calculated." This opened up a new series of representations of the "Woosung Bar" in the press. At the heart of this was a necessarily careful calculation, which demanded a further survey. I will call it a game on the hydrographic facts, which further exemplified the role of the press as a merchant propaganda machine in defining an environmental "crisis."

Mediating a game of hydrographical matter of facts

The suggestion made by the Foreign Ministers in Beijing had embarked another wave of media politics on the Woosung Bar. The hydrographic facts about the bar took center stage as different parties defended their positions. The discussion of the growth of traffic through Shanghai and the increasing size of the ships seemed to be obviated by the newspaper, while they shifted to discussing the "deteriorating" situation at Woosung. Their claims were in turn challenged by the Marine Department of Hart's Customs. The Marine Department served as the official hydrographical authority, collecting and maintaining data regarding the depth of the water on the bar since its establishment in 1868.[42] Adding to the existing tension between Hart and the Shanghai mercantile community caused by other issues, the representation in the press of the hydrographical matters of fact became critical.

In the report of the February 1872 survey, Viguier had defended the validity of the Custom's Department data by nuancing the way from which the sounding was made:

> In order to make the survey with greatest accuracy I had boats anchored in different parts of the river, and the lines of soundings run from one boat to another, and from the boats to stations on shore, the exact position of the boats and stations being determined by triangulation, and having erected a special tide-pole, close to the Harbour Master's Station, at Woosung, the register of the tides was kept with great care.

This detailed description was published in the *NCH*, forwarded by Seward (the U.S. Consul General in Shanghai). However, the result of this careful survey was still challenged by subsequent petitions. In the newspapers, the official data on the depth of water above the bar were represented as "erroneous" and blamed for ship detentions in the winter of 1871–1872.

The Marine Department of the Customs had kept a full series of data of river depth since its establishment. Prior to the establishment of the department in 1868, there seemed no systematic surveys of the bar. Yet, sporadic data were collected by British naval officers. Woosung River was surveyed by the British naval officers Commander John Ward in 1858 and Master E. Wilds in 1862. The result was the "Chart of Woosung River" published by the British Admiralty (Admiralty Chart, No. 1601).[43] The Chart was popular among ship masters and almost every future directory of navigation referred to it. The chart gave the depth of the water on the bar around 1860, recording a minimum depth of twelve feet during the spring tides. This was about what the data collected in the 1870s discovered, again suggesting that the newspaper's claims of a deterioration of the bar were false.

Even though frustrated by the diplomats in Beijing, the mercantile community of Shanghai turned abroad to voice their opinion on dredging. In 1872, two waves of energetic remonstrance of the Shanghai merchants had reached London. With the long-distance operations reaching "home," the shipping-insurance hegemony in treaty-port China demonstrated a great lobbying capacity. In April 1872, the British Foreign Office in London received a letter from J. A. Harper, the secretary of the Royal Exchange in London's Association for the Protection of Commercial Interest. In this letter, the communications regarding the bar between Morris, the Shanghai Chamber, and the Customs were presented before the British Foreign Office. The letter requested the "assistance of Her Majesty's Government, with a view to moving the authorities in Peking" to give their attention to the removal of a danger in the navigation of the port, which had then became the center of commerce in China.[44] A second London-bound petition was made by the P & O in September 1872, through the British General Post Office, asking that the "influence of Her Majesty's Government may be used to secure the attention of the Chinese Government to this important matter," since during the last few years the depth of water on this bar "has very much decreased."[45]

Receiving the first petition forwarded from London, Thomas Wade, the British ministerial "apologists in Peking," replied to his London superior, again explaining the Chinese official response. Robert Hart had been authorized by the Zongli Yamen to expend a considerable sum purchasing a dredging apparatus to be used at Woosung.[46] After the second petition, Wade and the British officers in the Foreign Office tried to obtain additional information regarding the situation at the bar. A new wave of circulation of hydrographical facts about the bar commenced. This time the discussion was out of the control of the Shanghai merchants and press. Wade had resorted to the British naval hydrographical authority in September 1872. The minister asked Admiral Charles F. A. Shadwell's help to have the question looked into by officers competent to report on it.[47] Shadwell was then commander in chief of the British Navy in the China Station. A report was accordingly made by Captain W. Arthur and Navigating Lieutenant Tracey of H.M.S. *Iron Duke*. Their report further undermined the narrative of deterioration. The report stated that "the Woosung Bar has not altered permanently in depth for many years."[48] The Shadwell concluded that the difficulty of navigation was "entirely owing to the increased size of the vessels at present employed in the trade to China."[49]

In addition to the statement of the British naval hydrographical authority, Hart's Customs also defended the position that the bar was basically unchanging. In February 1873, a table showing the daily depth of water on the Woosung Bar was unprecedentedly furnished to the *NCH* and occupied a whole page in an issue of the local paper (Table 4.3). The published table contained a series of records of daily high-water depth in the second half of 1872 and the average depth of each month during the same period. This table had very likely been furnished by customs officers, for numbers in it were the same as those in the meteorological tables carried regularly in the *NCH* and furnished by the customs station at Woosung. The table demonstrated again that there was almost no change in depth over the Bar compared to previous years. The seasonal variation of the bar was summarized at the bottom of the table.

To counter the Navy's testimony and the Customs' new evidence, the Shanghai press began to argue for the conservation of the Huangpu River, embracing a wider perspective. Although the U.S. and British ministers in Beijing argued that a thorough survey of the bar was necessary, the Shanghai press exonerated the merchants of such work. Instead, the *NCH* suggested that the Chinese government should take responsibility for the study, including not only the Woosong Bar but the nearby Yangtze River as well. A reliable survey, the newspapers pretentiously opined, could thus render "intelligent" operations that might prevent permanent injury. The would-be operations should obey the knowledge not only of "the general law of hydrostatics," but the particularities of the Yangtze.[50] The press then admitted that they had no necessary information on the situation and rate of silting on the bar, which was key to their argument in favor of dredging.[51]

Compared to the Customs' perennial tabulations on the bar, the information-deficient mercantile community continued their complaints about ship delays and detentions. The February 6, 1873 issue of the *NCH* had significantly exemplified the press as a field for the battle in this asymmetrical hydrographic battle. A letter to the editor in the issue by a reader "Truth," commented on the previously published table (Table 4.3), calling it misleading to anyone who had not "a practical knowledge" of the subject. The reasoning of "Truth" was that the tables were based on both the day and the night tides. But it was well known that the night tides, as a rule, are much higher than the day tides. Moreover, "Truth" argued, the high night tides were irrelevant for deep draft shipping, for they "were seldom able to get through the junks at Woosung during the night and never able to cross the Bar and come up the river at night without great risk." After his re-calculation taking only the daytime tides, the monthly average depth of the bar (Table 4.4) was much shallower. "Truth" concluded: "This fact speaks for itself, and shews [*sic*] that on the average of those 4 months there was not water of a vessel of 20 feet to cross the Bar."[52]

The facts of "Truth" were soon challenged by "Owl," who wrote another response the same week. "Owl" began: "as each tide is given … there is not much fear of the average depths misleading anyone." He suggested that "Truth" intentionally omitted the numbers for July and September, when the average a.m. high tides were more than twenty feet. "Owl" then accused "Truth" of being muddled about the a.m. and p.m. tides. "Truth" did not understand the real meaning of "day" and "night" in respect to tide data. "Owl" reminded readers that "a.m." referred to the period from midnight to noon and "p.m." from noon to midnight. After pointing more of "Truth" 's errors "Owl" concluded that he learned from one or two of the oldest pilots that they believed there was as much water on the bar now as there had been over ten years ago. Mocking the Shanghai merchants' lack of necessary information and defending the hydrographical authority of the customs' office, "Owl" continued: "a careful recording of the actual depth of water over a series of years is the only way to arrive at a proper understanding of this Bar, and such is now evidently being done by the Customs authorities."[53]

The appearance of an official summarize of water depths around Woosung Bar and the subsequent debate between "Truth" and "Owl" scarcely proved the press's neutrality in the debate. Customs was rarely in the favor of the local press. Most of the time, the voice of the commercial party was more prominent

Table 4.3 Depth of water on the Woosung Bar

Table shewing the depth of water on the Woosung Bar at each high water from July 1 to December 31, 1872

Day of the month	*July*				*August*				*September*				*October*				*November*				*December*			
	a.m.		*p.m.*		*a.m.*		*p.m.*		*a.m.*		*p.m.*		*a.m.*		*p.m.*		*a.m.*		*p.m.*		*a.m.*		*p.m.*	
	feet	*inches*	*feet*	*inches*	*feet*	*inches*	*feet*	*inches*	*feet*	*inches*	*feet*	*inches*	*feet*	*inches*	*feet*	*inches*	*feet*	*inches*	*feet*	*inches*	*feet*	*inches*	*feet*	*inches*
1	19	4	20	5	19	1	22	2	–	–	18	9	21	2	19	10	20	7	21	3	20	3	21	9
2	19	2	21	1	19	10	22	4	21	1	19	3	21	5	20	4	20	9	21	10	20	6	22	3
3	19	2	21	7	20	5	–	–	21	5	19	10	21	6	20	10	20	7	22	0	20	4	21	1
4	19	6	–	–	22	8	20	6	21	5	20	2	21	5	21	3	10	5	21	9	19	8	21	5
5	22	0	19	8	22	8	20	5	21	4	20	2	21	0	21	3	10	10	21	1	19	2	20	7
6	22	3	19	9	22	6	20	10	21	2	20	5	20	9	22	3	18	7	20	10	18	7	21	2
7	22	9	20	0	22	10	21	5	21	4	20	8	21	2	21	1	18	0	19	10	18	4	19	9
8	22	11	20	1	22	10	20	1	20	3	20	2	19	7	20	9	16	9	18	10	17	2	19	0
9	–	–	20	7	21	3	19	8	19	9	21	9	18	9	20	2	16	5	21	4	18	8	19	0
10	23	2	20	1	21	2	20	1	20	4	10	11	17	7	20	2	19	6	19	2	17	5	18	2
11	21	9	19	8	20	3	19	8	17	10	18	11	15	9	18	9	17	4	18	6	19	10	20	8
12	21	3	19	3	19	3	19	0	17	6	19	3	16	7	19	10	19	10	20	2	19	2	18	1
13	20	7	19	3	18	2	18	9	17	5	20	7	18	2	20	6	19	7	19	6	20	11	21	2
14	19	11	19	3	16	9	18	7	18	4	21	6	19	3	21	2	–	–	20	8	–	–	22	1
15	19	4	19	10	17	3	19	5	19	4	22	2	20	2	–	–	20	4	21	9	19	6	21	2
16	18	9	19	11	17	8	20	7	–	–	20	5	21	4	20	10	21	2	21	9	18	8	19	10
17	17	9	20	4	19	0	21	10	22	10	21	3	21	6	21	9	21	5	21	7	18	5	21	5
18	18	5	22	4	–	–	19	10	23	0	21	7	21	4	22	4	20	4	21	2	19	3	20	2
19	20	8	22	8	23	2	20	6	22	11	22	0	21	9	21	10	18	9	21`	2	17	11	20	8
20	–	–	19	11	22	9	21	0	22	4	22	2	21	6	21	9	18	7	20	8	18	7	20	2
21	23	4	20	9	22	7	21	1	21	8	21	11	20	10	21	2	18	2	20	4	17	7	19	5
22	23	7	21	2	22	7	21	2	20	9	21	5	19	3	20	3	17	7	20	0	17	5	19	1

23	23	11	21	3	22	2	20	10	19	6	20	10	18	1	19	3	19	3	10	3	17	4	19	0
24	23	7	21	2	21	8	20	10	18	4	19	10	16	5	18	10	16	0	19	3	17	4	18	4
25	23	2	20	8	20	3	20	3	17	5	18	10	15	8	17	8	16	10	17	11	17	11	18	1
26	21	9	19	8	18	7	19	2	15	8	18	2	15	5	18	2	17	5	19	1	19	3	18	2
27	20	3	19	7	16	8	18	5	15	10	18	9	16	8	18	8	19	5	19	4	19	4	18	6
28	19	3	19	5	15	11	18	3	17	1	19	7	18	1	19	6	19	4	10	3	10	11	10	1
29	18	4	19	2	16	1	19	1	18	9	20	7	19	10	20	10	20	2	–	–	–	–	22	5
30	17	8	19	3	16	10	19	6	19	4	–	–	20	0	20	7	20	3	20	11	19	10	21	4
31	17	8	21	3	17	7	19	11	–	–	–	–	–	–	20	5	–	–	–	–	18	10	22	10
Average depth of tide each month	*20 feet 6 inches*				*20 feet 1 inch*				*20 feet 1 inch*				*19 feet 0 inches*				*19 feet 7 inches*				*19 feet 6 inches*			

Number of tides of certain depths in each month

Month	*Tides from 15' to 15'6"*	*Tides from 15'6" to 16'*	*Tides from 16' to 16'6"*	*Tides from 16'6" to 17'*	*Tides from 17' to 17'6"*	*Tides from 17' 6" to 18'*	*Tides from 18' to 18'6"*	*Tides from 18'6" to 19'*	*Tides from 19' to 19'6"*	*Tides from 19'6" to 20'*	*Tides from 20' to 20'6"*	*Tides from 20'6" to 21'*	*Tides from 21 to 21'6"*	*Tides from 21'6" to 22'*	*Tides from 22' to 22'6"*	*Tides from 22'6" to 23'*	*Tides from 23' to 23'6"*	*Tides from 23'6" to 24'*
July	–	–	–	–	–	3	2	1	11	11	6	4	6	3	3	3	4	3
August	–	1	1	3	1	2	3	3	7	6	7	6	6	2	3	8	1	–
September	–	2	–	–	3	2	3	5	4	6	7	5	8	5	4	2	1	–
October	1	2	1	2	–	2	4	4	3	5	7	8	12	7	2	–	–	–
November	–	–	2	2	2	2	2	5	9	6	8	7	7	5	1	–	–	–
December	–	–	–	–	5	3	7	6	11	7	4	5	7	1	3	1	–	–
Totals	1	5	4	7	11	14	21	24	45	41	39	35	46	23	16	14	6	3

Source: *NCH*, February 6, 1873.

Table 4.4 Monthly average a.m. high tides of the bar calculated in a letter to the editor of the *NCH*

a.m. high water	*Feet and inches*	*a.m. high water*	*Feet and inches*
September	19 ft. 9 in.	November	18 ft. 9 in.
October	19 ft. 4 in.	December	18 ft. 10 in.

than that of the officials. From 1870 onward, cases of further detentions carried in the press had been typically titled "Woosung Bar, Again!"[54] Moreover, while the hydrographical matter of facts battle was taking place, on February 6, 1873, the same issue of the *NCH* carried more pages of another petition by mercantile interest to Beijing with full reprint of associated correspondence. In January, F. B. Johnson (Chairman of the Shanghai Chamber, partner in Jardine, Matheson & Co., and the Consul for Denmark sitting in the consular meetings of treaty powers in Shanghai) wrote to Wade, informing the British Minister in Beijing of the existence of the second petition to London. Johnson doubted if Hart had taken any of the actions he had promised, such as ordering a dredging machine in England. Johnson also asked what the Chinese government would do about the matter. He believed that the Chinese should grant the Shanghai mercantile community additional power to raise the funds themselves, "by means of special taxes to be levied upon native and foreign shipping resorting to the Port, and upon landed and house property within its limit."[55] The press accused that the delay (or more apparently non-action) on the dredging, had been more than official negligence. It was "a disgrace of the [Chinese] Government."[56]

Hart and his Customs, the constant registers of the bar and the "buffer" between the mercantile community and the Chinese government, had their own view. They were advisers rather than employees of the Chinese government and contributed to the policy of delay. The lack of sufficient information was their shared official excuse they sent Shanghai. In most occasions, Hart replied to the Shanghai merchants that he had no funds at his disposal for the dredging. Most of the tonnage dues were spent in lighting and buoying. The buffer seemed working for a delay when Hart promised that he would "urge the matter on the Chinese Government," only if he had necessary information. Hart's memorandum published in the *NCH*, further documented this position. In 1867, Hart wrote to the U.S. Minister in Beijing, informing that before any attempt be made with the bar

> a competent engineer will be brought from Holland, and on his opinion will depend the measures to be proposed … in the absence of the opinion of a man at once scientific and practical, who has made such works his study, my advice to the Chinese is to do nothing rather than to bury sycee in the mud.[57]

This echoed the same "lack-of-necessary-information" message sent to Shanghai by the British and U.S. ministers in 1872. To act upon the official inert reply, a survey was made by the Shanghai mercantile community themselves.

Mobilizing alternative hydrographic experts in Shanghai

In the winter of 1873, the mercantile community launched another petition with a new survey made in the summer, along with a cost estimate of dredging and compared to the costs incurred by delays caused by the Woosung Bar for different shipping companies. The survey was a collective response to the "suggestions" from Beijing. It aimed to demonstrate the feasibility of dredging and its manageable cost. F. B. Johnson, the Chairman of the Chamber of Commerce, employed the help of an American merchant, Charles E. Hill to survey the bar. Hill engaged Edward Hjousbery, a pilot in Shanghai, as superintendent of the survey. Johnson and Hill also asked the help of the Commander of the U.S. Naval Force on the Asiatic Station (Rear Admiral Thornton S. Jenkins), by resorting to the U.S. Consul General in Shanghai (Seward). With the order from the Admiral, the survey was made with the assistance of several U.S. naval officers (of the U.S.S. *Iroquois*). Hjousbery hereafter made a chart and gave a detailed report of how the survey was conducted, which was published in the *NCH*.[58] A dredging engineer named G. D. Hamill also sketched a scheme of dredging, and made the estimates of the number of cubic feet of mud that had to be removed in the work.[59]

The following meetings of the Shanghai Chamber and the Foreign Consuls discussed the new American survey. F. D. Barnes, the Agent of the P & O in Shanghai and latterly the Managing Director of the shipping giant's headquarters in London, presented on both occasions. The minutes of the chamber meeting were published in great detail by the *NCH*. Barnes collected various opinions from more than thirty pilots and captains. In Barnes' representation, almost all those local hydrographical experts had concurred in stating that the river was in general shoaling and narrowing. The bar had shoaled two feet during the previous eighteen months since 1872, when Hart was in Shanghai in person and promised action. Even worse, it seemed the water above the bar was two feet more shallow.[60] Recent detention cases owing to the bar occurred to P & O during the last two years were listed and laid before the meeting of consuls, in which the direct and indirect loss amounted to 22,000 taels since September 1871. Other major shipping companies also reported their losses. For instance, William Lang, of Messrs. Butterfield & Swire Company, testified that the Company lost 17,100 taels since January 1871. A. Hennequin added that in a single year the Messageries Maritimes suffered a loss of 33,400 taels due to the bar.[61]

Meanwhile, the Shanghai press had revised their rhetoric on the deterioration at the bar, but did so with a wider scope. They started to advocate a "Board of Conservation" to deal with not only the bar, but more comprehensively, other nuisance of the river, mainly wrecked ships and shoaling at other sites in the Huangpu. In October 1873, an editorial of the *NCH* admitted that

> it was a moot point whether or not the bar was really silting up, it might have been urged that the increased size of the steamers frequenting Shanghai was, at bottom, the origin of the complaints which have been urged of late years.

The same editorial argued: "facts are too strong to admit of this rejoinder, and for better or worse Shanghai must make up its mind to look upon silting of the Woosung Bar as a matter of fact."[62] The newspaper continued to point out that the bar was then not only detaining large ocean-going mail steamers, but also the smaller Yangtze navigating boats. The experience of the Shanghai Steam Navigation Company suggested that the situation was in "fact" worsening.

> It is only within the last year, we believe, that any of their boats have had occasion to lie at Woosung waiting for water to cross. Within the last two months, it has become necessary to arrange their time that they shall not arrive at Woosung at low water.

In short, the Bar was, the newspaper held, becoming more and more troublesome, and only did the local community perceive of it as "peculiarly a local matter.… No one knows where the shoe pinches so well as the sufferers—the Shanghai shipping interest."[63]

The 1873 local survey and the further revised complaints of the bar seemed unable to alter the minds of officials in Beijing. Another memorandum by Hart in the spring of 1874, in contrast to his earlier memo in 1867, was quite offensive to the Shanghai mercantile community. This confidential but latterly published document demonstrated the up-to-date hydrographical knowledge obtained by Hart and his Customs staff. In which the bar was considered as a part of the whole inter-related depositing phenomenon at the mouth of the Yangtze, which was not feasible to be solved by mere dredging. In the 1874 memorandum, Hart indicated that the agitation and repeated remonstrations from Shanghai were in fact a factor of commercial competition. He said:

> The trade consequent on opening the Yangtze River has so far been diverted into a false channel by the vested rights or money spent in Shanghai. This agency is in turn counteracted by the opening of the Suez Canal, through which steamers have begun to pass, making London and Hankow their termini. The tendency of the S. S. N. Company's steamers' operation is to bring teas to Shanghai; the competition of Holt's line [of the Butterflied & Swire] does not altogether oppose this tendency; but Holt's line, a combination of river and sea-going steamers, is being followed by other lines, and this competition will tend to foreign trade more and more to the steamers which do not call at Shanghai. Even the P & O preference for Shanghai will have to move on, and keeper competition will weaken the influence of Shanghai's priority and vested rights. Add to all, the competition coming from the Chinese side, which, in ten or twenty years' time, will have swept the foreign flag from the coasting trade of China, and displayed the Chinese colors in the London and Liverpool docks. The highway of trade for Central and Northern China will be the Yangtze. Teas will be shipped at Hankow and Kiukiang, and Shanghai silks and Ningpo teas at Chinkiang. They will be the return cargoes of the steamers which carry what China may continue

> to demand from Europe. In 20 years' time Chinkiang will have taken the place of Shanghai as a semi-terminus and trans-shipment port, but its foreign community will naturally be a smaller one, [political] possibilities aside. Thus looked at, as it affects and is affected by natural and artificial agencies now at work at the mouth of the Yangtze, the question of the Woosung Bar is seen to mean that dredging there may possibly be nothing more than a means of making the last days of Shanghai a little more comfortable than they would otherwise be; it will not prolong or avert the commercial death of the place, but it will make a show of vitality during its declining years more possible.[64]

This memorandum explained the official position on the Woosung Bar which denied any substantial steps as demanded by the Shanghai merchants. However, it was apparent that in the spring of 1874 Hart had overlooked the determination of the Shanghai mercantile community to protect their local interests.

Prior to Hart's second memorandum, the Shanghai merchants resumed petition with even greater efforts. In January 1874, a P & O petition again reached London. The collective survey of the bar in the winter of 1873 became additional evidence used to mobilize support in the home government. They had cried to the British government that "the state of the river is very much worse than it was some years since, and that, even within the last few months, it has deteriorated very considerably."[65] Though this seemingly urgent situation was discredited by the later paragraph of the same letter: "It is true that the [P & O] Company's homeward mail-steamers have not lately suffered any considerable detention; but this is due to the fact that their dates of departure have happened to fall on the spring tides." Again, this homeward petition was laid before the British Parliament in 1874. Previous correspondences between Shanghai, Beijing, and London regarding the bar between 1872 and 1874, were reprinted in a pamphlet for circulation among Parliament members.[66] The Shanghai press praised Robert Reid, an old China hand, for coming forward on behalf of the Shanghai merchants by sending question to Parliament. The Shanghai paper concluded: "it shows that [the British] Government is alive to the fact that there is a place called Shanghai, and is not altogether deaf to its requirements."[67]

The British Minister in Beijing had to be reluctant and accordingly forwarded the return petition from London to the Chinese government. Pressure came not only from the British Parliament. The Shanghai newspapers also noted that they were glad to hear from Beijing that the French Minister received from his government instructions identical to Wade's. Wade replied to London that "the matter has been pressed upon the Chinese Government for the last two years.... Since 13th of February, besides verbal communications, I have written three despatches to the Prince [Prince Kung 恭親王, the head of the Zongli Yamen]."[68] After receiving further official denials from the Chinese government, tensions soared. In the spring of 1874, the Chinese government was dealing with an international crisis, due to Japanese invasions of aboriginal villages in southern Taiwan. The "heavenly sent barrier" became a reality in the bellicose atmosphere when Prince Kung again denied any support for a dredging due to their urgent need to defend Woosung and protect the Huangpu River. Although the "local pressure" from Shanghai had been already sounding monotonously in the ear of the Beijing authorities, the Shanghai newspapers remained energetic and argued against Prince Kung's reply.[69] Hart summarized the situation in a telegraph dispatched to London. He wrote: "The Chinese refusal to allow the Woosung Bar to be dredged may become almost a *casus belli.*"[70]

Conclusion: media, commercial interests, and an environmental "crisis"

The commercial press in nineteenth-century Shanghai visualized a controversy on shipping access. The need for dredging the Woosung Bar had been raised, but the facts surrounding the issue were manipulated. Since 1870, the press had reported on most of the mercantile community's views and activities surrounding the bar, including the official responses of both foreign and Chinese authorities. The commercial papers highlighted the need for dredging as port traffic increased. Numerous petitions from the Shanghai General Chamber of Commerce and major shipping and insurance companies, were reported on and even enhanced by the newspapers. Through this the newspapers exerted "local pressure" upon the local Chinese magistrates, the Customs staff in Shanghai, and more distant authorities both in Beijing and "at Home." Newspapers also published various documents and correspondence, including the minutes of meetings of the Chamber of Commerce, the foreign consular meetings in Shanghai, correspondence between the Chamber and local magistrate, or between other officials and diplomats both in China and in London. These documents were both reprinted and discussed in the pages of the *NCH*. This profiling and documentation had enhanced the "public sphere" among the mercantile community and facilitated further collective actions, such as their autonomous survey in 1874. The press served as the merchants' propaganda machine and was far from a neutral party in the public discourse. The case of Woosung Bar had demonstrated the practice of this new and powerful "public" opinion in treaty-port China. Furthermore, it had greatly enhanced the commercial initiatives with their joint lobbying capacity. While the hydrographical facts in support of the merchants' dredging campaign were dubious, the manipulation was facilitated by the newspapers.

It is significant to our refreshed curiosity on the role of the media in the modern capitalist world. Environmental crises, such as extreme weather, are affecting people in everyday life, and have been reported enthusiastically in the news media. The image of these new crises might not be "false," as was the manufactured crisis of the Woosung Bar in 1870s Shanghai. But some lessons might be herewith undermined. The Shanghai press had never been a neutral forum for opinions of different sides in the controversy. The press's representation of the hydrographical data suggests a willingness to manipulate information delivered to the public. In contrast to the media-enhanced discourse of the bar's deterioration and persistent propaganda and public pressure, Robert Hart's and the Customs' collection of knowledge on the bar through surveys, such as the one done by the British Admiralty under the request of Thomas Wade, were virtually ignored by the newspapers. In contrast, the pro-merchant argument that conditions at the bar were worsening rapidly and the shouting of the large shipping companies prevailed in shaping public opinion in Shanghai.

Notes

1 Maxwell T. Boykoff, *Who Speaks for the Climate? Making Sense of Media Reporting on Climate Change* (Cambridge: Cambridge University Press, 2011), pp. 28–29.
2 Wu Yi-Xiong 吳義雄, "The Formation of a General Sentiment among Westerners in China for Waging War against China before the Opium War," *Journal of Modern History* [近代史研究] 2 (2009): 23–43
3 Christopher A. Reed, *Gutenberg in Shanghai: Chinese Print Capitalism, 1876–1937* (Vancouver: University of British Columbia Press, 2004); Barbara Mittler, *A Newspaper for China?: Power, Identity, and Change in Shanghai's News Media, 1872–1912* (Cambridge, MA: Harvard University Asia Center, 2004); Rudolf G. Wagner, ed., *Joining the Global Public: Word, Image, and City in Early Chinese Newspapers, 1870–1910* (Albany, N.Y.: State University of New York Press, 2007); Tsai Weipin 蔡維屏, *Reading Shenbao: Nationalism, Consumerism and Individuality in China, 1919–37* (New York: Palgrave Macmillan, 2010); Yong Z. Volz and Chin-chuan Lee, "Semi-colonialism and Journalistic Sphere of Influence," *Journalism Studies* 12:5 (2011): 559–574; for a most recent account, see Rudolf G. Wagner, "Don't Mind the Gap: The Foreign Press in Late-Qing and Republican China," *China Heritage Quarterly*, 2012, available at: http://chinaheritagenewsletter.anu.edu.au/features.php?searchterm=030_wagner.inc&issue=030, accessed on January 17, 2015.
4 An exception, though not focusing on the role of the press, might be the pioneering study by Chang Ning 張寧, "Cultural Conflicts in the British Community in China: Shanghai 'Sportsmen' vs. the 'Bird Slaughters,' 1890–1920 (在華英人間的文化衝突：上海 '運動家' 對抗 '鳥類屠害者', 1890–1920)," *Bulletin of the Institute of Modern History, Academia Sinica* [中央研究院近代史研究所集刊] 34 (2000): 89–144.
5 A recent example, see Niv Horesh, "Location Is (Not) Everything: Re-assessing Shanghai's Rise, 1840s–1860s," *Provincial China* 1.2 (2009): 61–75; for a classical account see H. B. Morse, *The International Relations of the Chinese Empire* (London: Longmans, Green & Co., 1910–1918), Vol. 2, pp. 275–276; Vol. 3, pp. 380–386.
6 John W. King, *The China Pilot*, 3rd ed. (London: Printed for the Hydrographic Office, Admiralty, 1861), p. 198.
7 Imperial Maritime Customs, *Woosung Inner Bar* (Shanghai: Kelly & Walsh, 1894), 6; Note on a map of "Wusung River or Huang Pu," made by A. M. Bisbee, assisted by R. Braun, 1887, in Imperial Maritime Customs, *Woosung Inner Bar*, 8a. Regarding the study of tides and its significance in the history of science, see Michael S. Reidy, *Tides of History: Ocean Science and Her Majesty's Navy* (Chicago: The University of Chicago Press, 2008).
8 Lindsay and Gützlaff, *Report of Proceedings on a Voyage to the Northern Ports of China in the Ship Lord Amherst* (London: B. Fellowes, Ludgate Street, 1833), pp. 169–170.
9 *History of Shanghai Harbor* [上海港史話], (Shanghai: Shanghai renmin chubanshe, 1979), p. 110.
10 *History of Shanghai Harbor*, pp. 111–112; "The Huangpu Conservancy Scheme," *NCH*, July 31, 1901; Francis Lister Hawks Pott, *A Short History of Shanghai* (Shanghai: Kelly & Walsh, 1928), pp. 100–101.

11 Imperial Maritime Customs, *Notice to Mariners, 1862–82* (Shanghai: Statistical Department of the Inspector General, 1883), Notices No. 1 (July 7, 1862) to No. 7 (February 23, 1863); Notice No. 5 (February 2, 1863) was a detailed "Directions for crossing the Inner Bar at Woosung" by the Harbor Master of Shanghai (Hockly).

12 Next to "Passengers" in the shipping news column of the *NCH*, June 24, 1865. Hart departed to Beijing through Tientsin by the steamer *Ying-tze-fei*.

13 For an example of this common suggestion, see "The Marine Department," *NCH*, December 13, 1870.

14 The establishment of the Marine Department in the Customs in 1868 was detailed by Chen Shi-qi 陳詩啟, "The Establishment of the Marine Department of the Customs and its Work in Modern China (中國近代海關海務部門的設立和海務工作的設施)," *Modern Chinese History Studies* [近代史研究] 6 (1986): 94–112.

15 "Editorial Selections: The River," *NCH*, October 25, 1871. For the full text of the treaty, see William Frederick Mayers, *Treaties between the Empire of China and Foreign Powers: Together with Regulations for the Conduct of Foreign Trade, Conventions, Agreements, Regulations, etc.* 4th ed. (Shanghai: North-China Herald Office, 1902), p. 16.

16 Robert Hart to James Duncan Campbell, May 16, 1874 in Xiafei Chen and Rongfang Han, (eds.), *Archives of China's Imperial Maritime Customs: Confidential Correspondence Between Robert Hart and James Duncan Campbell, 1874–1907* (Beijing: Foreign Language Press, 1990), Vol. 1, pp. 18–19.

17 "The Marine Department," *NCH*, December 13, 1870.

18 The "shipping-insurance hegemony" also exerted their great influence on the incipient public weather service in Shanghai and Hong Kong. See Marlon Zhu, "Typhoons, Meteorological Intelligence and the Inter-Port Mercantile Community in Nineteenth-century China," (Ph.D. diss., State University of New York, Binghamton, 2012).

19 For the discussion of the steam shipping industry, see Kwang-Ching Liu, *Anglo-American Steamship Rivalry in China, 1862–1874* (Cambridge, MA: Harvard University Press, 1962); Anne Reinhardt, "Navigating Imperialism in China: Steamship, Semicolony, and Nation, 1860–1937," (Ph.D. diss., Princeton University, 2002).

20 "Trade with China," *Times* (London), September 7, 1864; despatch, Sir Frederick Bruce (British Minister in China in Beijing) to British Foreign Office, June 7, 1864.

21 *Times* (London), September 12, 1864; partially reprinted in "China According to the *Times*," *NCH*, November 26, 1864.

22 British Consular Report, 1867, Shanghai, 117–118.

23 Yen-ping Hao, *The Commercial Revolution in Nineteenth-Century China: The Rise of Sino-Western Capitalism* (Berkeley: University of California Press, 1986); Nie Baozhang 聶寶璋, "The Expansion of the Foreign Firms in China in the 1860s (19世紀60年代外國在華洋行勢力的擴張)," *Selected Works of the Scholars in the Chinese Academy of Social Sciences: Volume of Nie Baozhang* (Beijing: Zhonghguo shehui kexue chubanshe, 2002), pp. 43–78, originally published in *Historical Research* [歷史研究] 6 (1984).

24 Morse, *The International Relations of the Chinese Empire*, Vol. 2, p. 275.

25 "The Chamber of Commerce," *NCH*, November 22, 1870; *NCH*, March 1, 1871. Regarding the intermediate role of Taotai between foreigners and Chinese higher authorities, see Yuan-sheng Liang, *The Shanghai Taotai: Linkage Man in a Changing Society, 1843–90* (Honolulu: University of Hawaii Press, 1990).

26 *NCH*, March 1, 1871; published correspondence, Viguier to Thomas Dick (Commissioner of Customs in Shanghai), February 11, 1871, *NCH*, March 8, 1871. This letter was transmitted to the *NCH* by George F. Seward, Senior Consul of the Consul Body and U.S. Consul General in Shanghai. Those international naval officers who verified Viguier's survey were H. Koch (First Lieutenant, Denmark), D. G. Davidson (Commander, Britain), Vincent (Lieutenant, France), and E. P. McCrea (Commander, U.S.).

27 For an introduction of P & O, see Freda Harcourt, *Flagship of Imperialism: The P. & O. Company and the Politics of Empire from Its Origins to 1867* (Manchester: Manchester University Press, 2006).
28 Published correspondence, S. A. Viguier to Jellicoe, May 6, 1871, *NCH*, October 18, 1871.
29 *NCH*, October 11, 1871.
30 Published correspondence, Jellicoe to the Shanghai Chamber of Commerce, September 11, 1871, *NCH*, October 18, 1871.
31 Published correspondence Varagot to Ch. De Crety, September 12, 1871, *NCH*, October 18, 1871. The survey was made on September 11, 1871.
32 Published correspondence, Water Pearson (Secretary of the Chamber) to G. F. Seward (Senior Consul of the Consular Body and the U.S. Consul General in Shanghai), *NCH*, October 18, 1871.
33 "Meeting of the Treaty Consuls," *NCH*, October 18, 1871.
34 Morris held this post since no later than 1868 and retired from the post on December 31, 1896. See *NCH*, April 23, 1897. He was once "District Grand Treasurer" of the Freemasonry in Shanghai in 1881–1882. See Robert S. Ivy, *The History of Freemasonry in Shanghai and North China* (Tientsin: The North China Printing and Publishing Company, 1913), 36.
35 Published correspondence, Herbert S. Morris and seventy-six other firms and insurance companies to Dick, 11 January, 1872, *NCH*, February 1, 1872; this letter is also in the report of Seward to the U.S. Secretary of State, No. 521, February 21, 1872, which document contains the full list of the co-signers. On page 88 of the issue of the *NCH*, the number of signers was counted as seventy-nine.
36 Robert I. Fearon (Vice-Chairman of the Shanghai Chamber of Commerce) to the Ministers of Treaty Powers in Beijing, February 16, 1872, in Seward (U.S. Consul General in Shanghai) to the U.S. (Acting) Secretary of State, United States National Archives, Washington, D.C., "MS Despatches from the U.S. Consuls in Shanghai, China, 1847–1906," Vol. 16, No. 521, February 21, 1872.
37 Published correspondence, Low to Seward, February 21, 1872, *NCH*, March 21, 1872.
38 *NCH*, March 21, 1872.
39 A complaint of the delay, see the "What People are Saying," *NCH*, March 7, 1872.
40 Published correspondence, Wade to Robert I. Fearon (Vice-Chairman of the Shanghai Chamber of Commerce), March 26, 1872, *NCH*, April 11, 1872.
41 "The Woosung Bar," *NCH*, March 28, 1872. For a similar criticism both on the Chinese government and the Foreign Ministers in Beijing, see "What People are Saying," *NCH*, May 25, 1872, which reads "Chinese obstructiveness [*sic*], backed by Ministerial obtuseness, is too much for anybody."
42 At the beginning of the establishment of the Department, the Shanghai press had correctly considered it as "a large Hydrographic Department" to be organized by the Customs, see *NCH*, October 27, 1868.
43 Imperial Maritime Customs, *Woosung Inner Bar* (Shanghai: Kelly & Walsh, 1894), p. 2.
44 J. A. W. Harper (Secretary of the Association) to E. Hammond (British Foreign Office), April 18, 1872 in *Correspondence Respecting the State of the Woosung Bar, near Shanghae, Presented to both Houses of Parliament by Command of Her Majesty, 1874* (London: Harrison and Sons, 1874), No. 1.
45 A. M. Bethune (Secretary of the P & O in London) to F. Mill (Secretary to the British Postmaster-General), September 14, 1872, in *Correspondence*, No. 4.
46 Wade to Earl Granville (British Secretary for Foreign Affair), July 22, 1872, in *Correspondence*, No. 7.
47 Wade to Earl Granville, December 5, 1872, in *Correspondence*, No. 12.
48 Report by Captain Arthur, December 1, 1872, in *Correspondence*, Enclosure 2 of the letter No. 12.

49 Shadwell to Mr. Wade, January 14, 1873, in *Correspondence*, Enclosure 1 of the letter No. 14.
50 "Changes in the Yangtze," *NCH*, February 6, 1873.
51 "The Woosung Bar," *NCH*, February 6, 1873.
52 "Truth," letter to the editor, *NCH*, February 6, 1873.
53 "Owl," letter to the editor, *NCH*, February 6, 1873.
54 For instances, see *NCH*, January 4, 1872 and *NCH*, February 1, 1872.
55 Published correspondence, Johnson to Wade, January 20, 1873, *NCH*, February 6, 1873. Pott, *A Short History of Shanghai*, p. 101, had wrongly dated Johnson's petition in 1864. The suggestion of the additional tax was declined by Wade.
56 "Changes in the Yangtze," *NCH*, February 6, 1873.
57 Hart's memorandum (January 31, 1867) enclosed in a letter from Anson Burlingame (U.S. Minister in Beijing) to Seward (U.S. Consul General in Shanghai), March 29, 1867, *NCH*, June 5, 1875.
58 Published correspondence, Hjousbery to Hill, November 14, 1873, *NCH*, November 27, 1873. The Chamber paid for expense of the survey with $376.82.
59 Seward to the U.S. (Assistant) Secretary of State, United States National Archives, Washington, D.C. "MS Despatches from the U.S. Consuls in Shanghai, China, 1847–1906," Vol. 16, No. 677, December 11, 1873, Documents laid before the Body of Foreign Consuls at Shanghai in relation to the Harbor of Shanghai, the Woosung Bar, and the Whangpoo River.
60 *NCH*, November 27, 1873; *NCH* December 4, 1873; *NCH* March 18, 1880.
61 *NCH*, November 27, 1873; the detailed calculation of the former two companies, see Seward to the U.S. (Assistant) Secretary of State, United States National Archives, Washington, D.C. "MS Despatches from the U.S. Consuls in Shanghai, China, 1847–1906," Vol. 16, No. 677, December 11, 1873, pp. 8–9.
62 "The Woosung Bar," *NCH*, October 9, 1873.
63 "The Woosung Bar," *NCH*, October 9, 1873.
64 "Hart's Memorandum (March 5, 1874)," *NCH*, April 24, 1875. Part of these paragraphs were also cited by Pott, *A Short History of Shanghai.*, pp. 101–102; and Rhoads Murphey, *Shanghai: Key to Modern China* (Cambridge, MA: Harvard University Press, 1953), pp. 75–76.
65 R. M. Bethune (Secretary of the P & O, London) to John Tilley (of General Post Office, London), January 20, 1874, *Correspondence Respecting the State of the Woosung Bar, near Shanghae, Presented to both Houses of Parliament by Command of Her Majesty, 1874* (London: Harrison and Sons, 1874), Enclosure in No. 17.
66 *Correspondence Respecting the State of the Woosung Bar*. This pamphlet was also reprinted by the *NCH* of July 4, 1874.
67 *NCH*, May 2, 1874.
68 Wade to the Earl of Derby (telegraph), April 3, 1874, *Correspondence Respecting the State of the Woosung Bar*, No. 20.
69 Both Prince Kung's reply and the following comment of the Press, see *NCH*, June 27, 1874.
70 Hart to Campbell (telegram), May 16, 1874, Xiafei Chen and Rongfang Han eds., *Archives of China's Imperial Maritime Customs: Confidential Correspondence Between Robert Hart and James Duncan Campbell, 1874–1907* (Beijing: Foreign Language Press, 1990), Vol. 1, pp. 18–19.

Bibliography

Boykoff, Maxwell T. *Who Speaks for the Climate? Making Sense of Media Reporting on Climate Change*. Cambridge, UK: Cambridge University Press, 2011.

British Consular Report, 1867, Shanghai.

Chang Ning 張寧. "Zaihua yingren jian de wenhua chongtu: Shanghai 'yundon jia' duikang 'niaolei duhai zhe' 1890–1920 在華英人間的文化衝突：上海 '運動家'對抗 '鳥類屠害者,' 1890–1920 (Cultural Conflicts in the British Community in China: Shanghai 'Sportsmen' vs. the 'Bird Slaughterers,' 1890–1920)," *Zhongyang yanjiuyuan jindaishi yanjiu jikan* 中央研究院近代史研究所集刊 [*Bulletin of the Institute of Modern History, Academia Sinica*] 34 (2000): 89–144.

Chen Shi-qi 陳詩啟. "Zhonguo Zhindai Haiguan Haiwu Bumen de Sheli he Haiwu Gonzhuo de Sheshi 中國近代海關海務部門的設立和海務工作的設施 (The Establishment of the Marine Department of the Customs and its Work in Modern China)," *Jindaishi Yanjiu* 近代史研究 [*Modern Chinese History Studies*] 6 (1986): 94–112.

Chen, Xiafei and Rongfang Han, eds. *Archives of China's Imperial Maritime Customs: Confidential Correspondence Between Robert Hart and James Duncan Campbell, 1874–1907*. Beijing: Foreign Language Press, 1990.

Correspondence Respecting the State of the Woosung Bar, near Shanghae, Presented to both Houses of Parliament by Command of Her Majesty, 1874. Number 1. London: Harrison and Sons, 1874.

Hao, Yen-ping. *The Commercial Revolution in Nineteenth-Century China: The Rise of Sino-Western Capitalism.* Berkeley: University of California Press, 1986.

Harcourt, Freda. *Flagship of Imperialism: The P. & O. Company and the Politics of Empire from its Origins to 1867*. Manchester: Manchester University Press, 2006.

Horesh, Niv. "Location Is (Not) Everything: Re-assessing Shanghai's Rise, 1840s–1860s," *Provincial China* 1.2 (2009): 61–75.

Imperial Maritime Customs. *Notice to Mariners, 1862–82*, Shanghai: Statistical Department of the Inspector General, 1883.

Imperial Maritime Customs. *Woosung Inner Bar*. Shanghai: Kelly & Walsh, 1894.

Ivy, Robert S. *The History of Freemasonry in Shanghai and North China*. Tientsin: The North China Printing and Publishing Co., 1913.

King, Frank H. H. and Prescott Clarke, *A Research Guide to China-coast Newspapers, 1822–1911*. Cambridge: Harvard University Press, 1965.

King, John W. (Master, RN). *The China Pilot*. 3rd ed. London: Printed for the Hydrographic Office, Admiralty, 1861.

Liang, Yuan-sheng. *The Shanghai Taotai: Linkage Man in a Changing Society, 1843–90*. Honolulu: University of Hawaii Press, 1990.

Lindsay and Gutzlaff. *Report of Proceedings on a Voyage to the Northern Ports of China in the Ship Lord Amherst*. London: B. Fellowes, Ludgate Street, 1833.

Liu, Kwang-Ching. *Anglo-American Steamship Rivalry in China, 1862–1874*. Cambridge, MA: Harvard University Press, 1962.

Lubot, Eugene. "Modern Chinese History as Reflected in the *North China Herald*," *The Social Studies* 65:3 (1874): 126–129.

Mayers, William Frederick. *Treaties between the Empire of China and Foreign Powers: Together with Regulations for the Conduct of Foreign Trade, Conventions, Agreements, Regulations, etc.* 4th ed. Shanghai: North-China Herald Office, 1902.

Mittler, Barbara. *A Newspaper for China?: Power, Identity, and Change in Shanghai's News Media, 1872–1912*. Cambridge, MA: Harvard University Asia Center, 2004.

Morse, Hosea Ballou. *The International Relations of the Chinese Empire*. 3 vols. London: Longmans, Green & Co., 1910–1918.

Murphey, Rhoads. *Shanghai: Key to Modern China*. Cambridge, MA: Harvard University Press, 1953.

Nie, Baozhang 聶寶璋. "19 shiji 60 niandai waiguo zaihua yanghang shili de kuozhang

19世紀60年代外國在華洋行勢力的擴張 (The Expansion of the Foreign Firms in China in the 1860s)," in *Selected Works of the Scholars in the Chinese Academy of Social Sciences: Volume of Nie Baozhang*. Beijing: China Social Science Press 中國社會科學出版社, 2002, pp. 43–78.

North China Herald.

Pott, Francis Lister Hawks. *A Short History of Shanghai*. Shanghai: Kelly & Walsh, 1928.

Reed, Christopher A. *Gutenberg in Shanghai: Chinese Print Capitalism, 1876–1937*. Vancouver: University of British Columbia Press, 2004.

Reidy, Michael S. *Tides of History: Ocean Science and Her Majesty's Navy*. Chicago: The University of Chicago Press, 2008.

Reinhardt, Anne. "Navigating Imperialism in China: Steamship, Semicolony, and Nation, 1860–1937," Ph. D. diss., Princeton University, 2002.

Shanghai Gang Shihua 上海港史話 [*A History of Shanghai Harbor*]. Shanghai: Shanghai ren min chu ban she 上海人民出版社, 1979.

Times (London).

Tsai, Weipin. *Reading Shenbao: Nationalism, Consumerism and Individuality in China, 1919–37*. New York: Palgrave Macmillan, 2010.

Volz, Yong Z. and Chin-chuan Lee. "Semi-colonialism and Journalistic Sphere of Influence," *Journalism Studies* 12:5 (2011): 559–574.

Wagner, Rudolf G. ed. *Joining the Global Public: Word, Image, and City in Early Chinese Newspapers, 1870–1910*. Albany: State University of New York Press, 2007.

Wagner, Rudolf G. "Don't Mind the Gap: The Foreign Press in Late-Qing and Republican China," *China Heritage Quarterly*, 2012, available at: http://chinaheritagenewsletter.anu.edu.au/features.php?searchterm=030_ wagner. inc&issue=030 (accessed on January 17, 2015).

Wu, Yi-Xiong 吳義雄. "The Formation of a General Sentiment among Westerners in China for Waging War against China before the Opium War," *Jindaishi Yanjiu* 近代史研究 [*Journal of Modern History*] 2 (2009): 23–43.

Zhu, Marlon. "Typhoons, Meteorological Intelligence and the Inter-Port Mercantile Community in Nineteenth-century China," Ph.D. diss., State University of New York, Binghamton, 2012.

5 Disputes on the expansion project of Dahaizi Reservoir, Xinjiang, China, in the 1950s

Li Zhang and Dongkai Yan

Introduction

Due to the changes in population structure, fuel-use, technology, economic growth models, lifestyles and politics, the environmental history of the twentieth century was completely different from the earlier times.[1] Human beings altered the hydrosphere as never before by constructing dams and reservoirs.[2] Many scholars have discussed the role of large water projects in modern Chinese environmental history.[3] Water project construction campaigns in China during the 1950s-1960s deserve more attention for they not only transformed the groundwater network but also played an important role in nation-building and economic development. Thus the making of Chinese environmental history was different from previous years, and it was also an important transition period.

After the peaceful political power transition in Xinjiang at the end of 1949, frontier consolidation became the first core issue next only to economic development for the central and local government. There were about 191,000 soldiers in Xinjiang. How to supply these soldiers with enough food was a serious issue. The most serious food-shortage area in 1950 was the region around Dihua (迪化), later named Urumqi (烏魯木齊), the capital of Xinjiang. The cost of grain transportation from interior China to Urumqi was very high; the cost of transporting 1 *dan* (石, 1 *dan* = 1 hectoliter) of grain would be 8 *dan*.[4] Therefore, 110,000 local soldiers were assigned to reclaim 400–470,000 ha of wastelands and to plant grain and cotton in January of 1950.[5] The Manas River (瑪納斯河) Basin, with huge areas of uncultivated land and a large quantity of water resources, was regarded as the most suitable area for cultivation near Dihua.

"No water, no farmland" was an old saying in Xinjiang. It is a very dry region in Northwest China. Irrigation was a fairly significant issue for the exploration plan on the Manas River Basin. In order to control flood and to adjust uneven water distribution in different seasons, constructing big dams that create reservoirs, not water pools in the traditional way, was a new water management measure in the 1950s.[6]

Dahaizi (大海子) Reservoir was planned for construction in the early 1950s, and its earth dam, eight kilometers long and 16 meters high, was completed in 1952. It was the first reservoir in the plain of the Manas River Basin. In 1954, the local government put forward a plan to expand the reservoir, but Soviet experts strongly opposed it and suggested abandoning the original reservoir immediately. However, in 1958, local authorities decided to expand it and finished the project within a year. It was a very tortuous construction process for Dahaizi Reservoir. The process involved many unique features of Chinese history in the 1950s: land reclamation and frontier consolidation in Xinjiang, domestic situation changes in China, changes of Sino-Soviet relationship, changing attitudes toward the impacts of water project and so on.

What is perhaps distinctive in this case is the highly historical detail of the debate on reconstruction. Why did the expansion project of Dahaizi Reservoir raise the attention of the Soviet experts? What was the core concern of two sides in the debate? What factors in the debate had influenced the fate of Dahaizi Reservoir? What was the modernity mode of China in the 1950s?

Land reclamation, cotton production and the water resources planning of the Manas River Basin

Political power in Xinjiang was transferred peacefully at the end of 1949. There were about 191,000 Xinjiang garrison troops. In order to feed themselves as quickly as they could, Mao Zedong (毛澤東, 1893–1976) commanded Xinjiang troops to engage in agriculture near their camp on December 5, 1949. Then, 24 days later, Wang Zhen (王震, 1908–1993), the Commander-in-chief of the Xinjiang army, put forward a plan of farming: 100,000–110,000 soldiers (out of a total of 191,000) would reclaim and cultivate about 400,000–470,000 ha of lands.[7] Then they began to look for cultivable land and available water resources.

In March 1950, the twenty-fifth and twenty-sixth divisions of the Twenty-second Corps (二十二兵團) marched to the Manas River Basin, located in the northern foothills of the Tianshan (天山) Mountains in Xinjiang. Its area is about 2.29 million hectares. There are six rivers with a total annual runoff of 2.27 billion cubic meters of which the Manas River is the biggest, with runoff of 1.28 billion cubic meters. The Manas River is a typical mountain stream and is mainly supplied by snow-melt water. Its maximum flow occurs from June to August, and then it becomes a swift-flowing river. Therefore, it is very difficult to draw water from this river. Before 1950, there were some temporary dams in the river built by fine willow branches, reed and grass which were usually fragile and easily destroyed by summer floods. Meanwhile, the middle part of Manas River Basin is an alluvial plain with gradient of 1–3 percent and vegetation coverage of 15–20 percent (see Figure 5.1 and Figure 5.2). It is a fertile flood plain. At the end of 1949, there were only 59,000 people distributed across the lower part of the alluvial fans and dry delta.[8] There were huge areas of uncultivated land on the wide floodplain because there was no technology to utilize the water resources.

Wang Zhen considered that the soil of Manas River Basin was better than that of Nanniwan (南泥灣), where he made his soldiers reclaim wasteland from 1941 to 1945 for self-sufficiency in grain.[9] Therefore, in the early 1950s, the

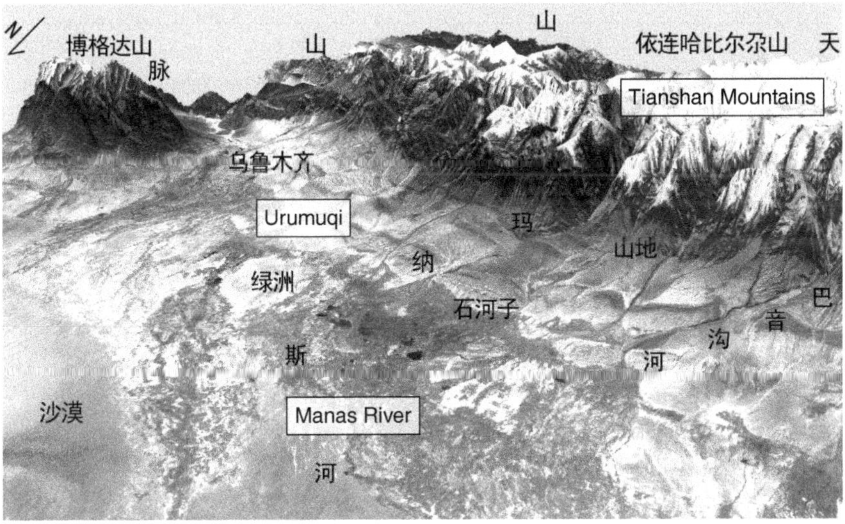

Figure 5.1 Three-dimensional map of the Manas River Basin in the northern slope of the Tianshan Mountains (source: Chen Weiming et al., "Manas he liuyu wushi nian luzhou kuozhan ji shengtai huanjing yanbian yanjiu (Oasis Expansion and Eco-environmental Evolution of Manas River Basin in the Past 50 Years)," *Zhongguo Kexue D: Diqiu Kexue (Science in China Series D: Earth Sciences)*, Vol. 35, No. 11 (2005): 1074–1086. (With permission from the author)).

Twenty-second Corps was asked to reclaim 13,300 ha of land. But in the winter of 1951, Wang Zhen ordered the troops to reclaim 33,300 ha of land to the west of the Manas River and ordered the Department of Water Resources of Xinjiang Province to cooperate with them.[10] The main purpose was that they would plant in spring and could be self-sufficient in grain at the end of that year. They must use new methods to provide enough water for these reclaimed lands. In order to plant as quickly as they could, they simultaneously repaired the old channels in Manas River Basin and surveyed and planned new channels.

In October 1951, Tao Zhiyue (陶峙岳, 1892–1988), the Vice Commander of Xinjiang Military Region and Commander of the Twenty-second Corps, and Wang Heting (王鶴亭, 1910–1996), Chief of the Water Conservancy Bureau of Xinjiang Province and also a famous hydraulic expert, together with more than ten other experts surveyed the Manas River Basin carefully. They realized that if they wanted to enlarge farmland they must take all measures to store winter water and flood water in reservoirs for use during the agricultural season. This was the guiding principle of the *Regional Planning of Manasi River and Kuitun River* made in May 1952:

> The Manas River Basin is the best large-scale agriculture area in northern Xinjiang near Dihua. It is one of the most important areas of irrigation

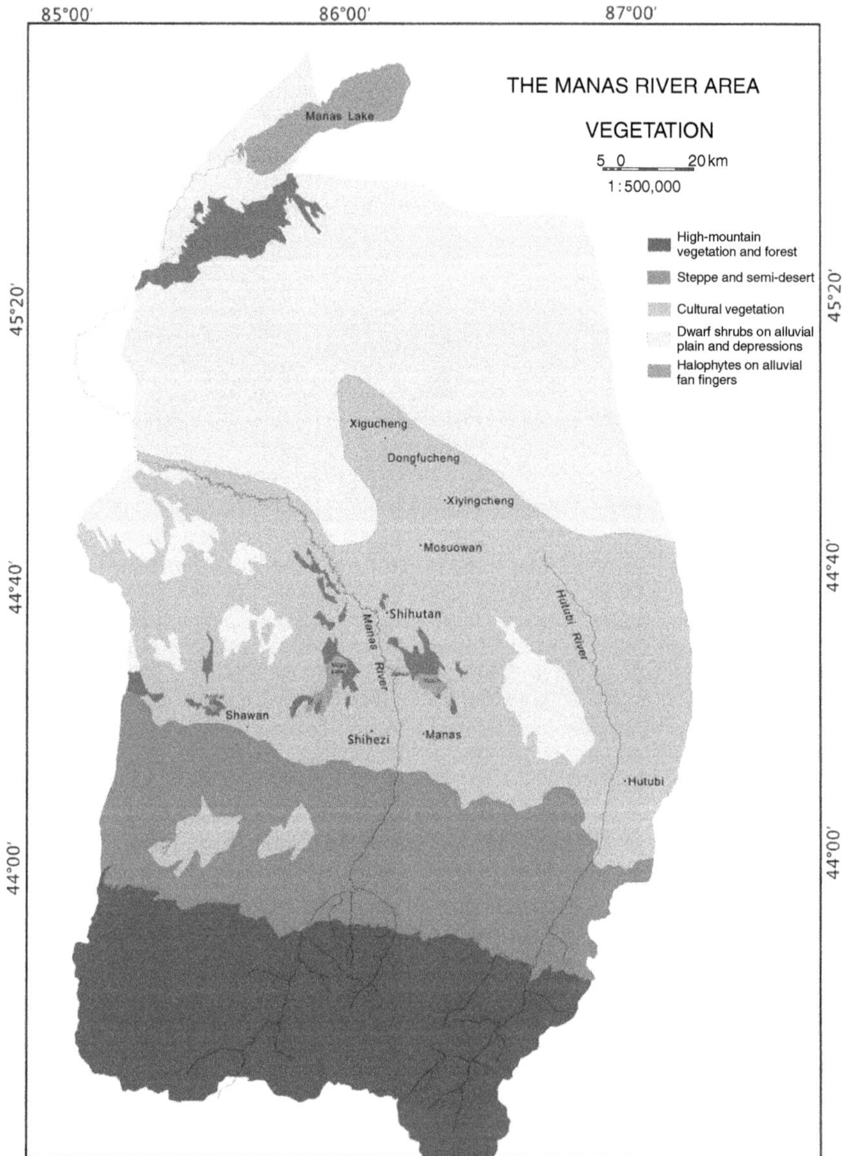

Figure 5.2 Simplified vegetation distribution in the Manas River area (source: This figure is drawn by the author (Li Zhang) based on Interpretation map from remote sensing image (ETM, August, 2000)).

> works construction in Xinjiang although there is not a very large amount of water resources. According to Wang Zhen's opinion, Commander of the Xinjiang Military Region, people should make full use of all water resources and must reclaim land at least 3.5 million *mu* (233,300 ha), then run up to 5 million *mu* (333,300 hectares) later. It is just a typical case of the ongoing Plan for the Transformation of Nature. So it is very critical to develop irrigation by constructing reservoirs to store snow-melt water in the spring for the crop needs in May and June, and store flood water in the summer. Then more land can be cultivated to establish large scale state-owned farms or collective farms for agricultural development. (瑪納斯河流域為北疆鄰近迪化最好的大農業生產地區，因此水源雖然不大，在新疆水利建設上仍列為重要水利中心之一．依照新疆軍區王司令員的指示，在這區域內將充分利用一切水源，必須開墾350萬畝地，爭取開墾500萬畝，並造林300萬畝，這是新疆正在進行的改造自然大計劃的一個典型部份．根據上述情況，這個地區今後發展灌溉的關鍵．主要在於辦理蓄水，必須將冬季不用的水源，春季大部份消冰水和夏秋一部分洪水儲蓄起來，解決五、六兩月的缺水，然後和七、八月間洪水結合起來經濟適用，就可以再大量擴充耕地，建立大規模的國營或集體農場，發展農牧業.)[11]

In this guiding principle, water and land were considered as natural resources and it was believed that humans should and could exploit them. This reflects the dominant approach toward water and uncultivated land at that time in China. The influence of the Soviet Union economic development model can be seen clearly in this planning. The Plan for the Transformation of Nature was proposed by Joseph Stalin (1878–1953) in the Soviet Union and was decided upon on October 20, 1948 for land development, agricultural practices and water projects to improve agriculture.[12] Constructing reservoirs to promote irrigated agriculture was one of the three main measures of this plan. The other two measures, planting trees and rotating crops and pasture in farmland, were also mentioned in this regional planning. Obviously, it was a Soviet-style planning of large-scale irrigation projects.

In the regional planning, they planned to construct four reservoirs and a big channel to draw river water to the huge flood plain. Two reservoirs were built at the west of Manas River, one at the east, and the last one in the river course and across the river. These hydraulic projects were reported in newspapers in October 1952 as a symbol of modernization.[13]

According to the guiding principle, people began to build an earth dam around a relatively low area on the east bank of Manas River to store the snow-melt water in the spring. It took only two months to construct an earth dam eight kilometers long and 16 meters high and create a plain reservoir, named Dahaizi Reservoir, with a capacity of four million cubic meters that could irrigate 533 hectares of land.[14] Its irrigation area had been more than the total farmland in the whole Manas River Basin.[15] However, it should be noticed that the project was as simple as it was in order to irrigate the new reclaimed lands by troops immediately. It used little capital but large amounts of labor and fairly simple technology. There were 17,000 laborers who participated in the dam construction with the help of a few traction machines and sheep-horn rollers in the traditional method. Besides, there was no careful hydrogeological drilling on the dam foundation beforehand. So shortly after the completion, the dam foundation seepage occurred. An expert criticized such hydraulic projects and expressed his doubts about their effectiveness.[16] However, most of the people at that time approved such new water utilization methods for the urgent irrigation needs in the dry area.

Two other important events occurred at the same time which closely related to the fate of Dahaizi Reservoir. One was the successful cotton planting in the Manas River Basin, where no cotton had been planted before. It was generally believed that 44 degrees of north latitude was the northern boundary of cotton growth. With the help of Soviet experts in 1950, cotton was planted experimentally in the Manas River Basin, located at about 45 degrees of north latitude. The average yield of cotton per *mu* (畝, 1 *mu*=666.67 square meters) increased year by year; it was 51.2 *jin* (斤, 1 *jin*=0.5 kilogram) in 1950, 78.9 *jin* in 1951, 120.4 *jin* in 1952 and 355.3 *jin* in 1953. It was also recorded that the yield of cotton reached 1,349 *jin* per *mu* in a special and small plot.[17] Cotton was a strategically important crop for Chinese industrialization and played a vital role in ordinary life in China and around the world.

Before 1949, more than half of the cotton consumption in China depended on imports, and the average yield of cotton per *mu* was about 200 *jin*. The successful cotton cultivation in the Manas River Basin and the high average yield immediately caught the attention of the whole country. On December 25 1953, *People's Daily* [*Renmin ribao* 人民日報] reported this with a title of "A Large Area of Cotton in the Manas River Basin Got Good Harvest," and pointed out that it was the highest average yield in the whole country, and there was strong evidence for the Manas River Basin to have great potentials to increase cotton production.[18] In the following two years, there were many reports about successes in cotton cultivation in the Manas River Basin.[19] Obviously, the Manas River Basin was well-known as a representative of modern agricultural areas and was heavily promoted as a symbol of national achievement. Therefore, the local and central governments had paid more attention to the reclamation and irrigation works.

Another important event was the Big West Channel (Xiandaqu 西岸大渠) project being halted by the central government in the winter of 1952. Similar to the principle of the above-mentioned regional planning, the Big West Channel was designed by local authorities and was constructed in early May of 1952 in order to draw water from the Manas River to the newly found 66,700 ha of uncultivated arable land known as Xiayedi (下野地) for agricultural cultivation. With the gradual nation-building of the new China, the central government began to standardize the financial approval procedures in Xinjiang in the second half of 1952. The Big West Channel project was criticized by the central government for it was not following the construction approval procedures and started construction before receiving the central government's approval. Therefore, the project was asked to stop and required to re-submit the application for approval, together with other water projects in the Manas River Basin. Directly, this event led to the beginning of the standardized and modern water resources plan of the Manas River Basin. At that time, the regular basin-wide plans of all big rivers had just begun to be drafted under the guidance of Soviet hydraulic experts.

Soviet experts opposed to expanding the Dahaizi Reservoir

The Manas River Basin had been very famous and was paid close attention by local and central governments. In June 1953, Su Zongsong (粟宗嵩, 1910–2009), a well-trained hydraulic engineer and the Chief of Irrigation Division of the Department of the Water Resources and Hydropower of the PRC, was assigned by the Department to go to Manas when the local government requested assistance. He was responsible for the final approval of water projects.[20] He led a team of more than ten hydraulic engineers to survey the whole river basin carefully and gave much good advice to revise the regional plan. The draft of *Water Resources Plan of the Manas River Basin* was completed in early 1954. In this plan, construction of reservoirs on the floodplain to restore almost all water resources was also the central part. Dahaizi Reservoir was one of the four reservoirs on the plain and was designed to expand to provide more water for more arable lands, located in Mosuowan (莫索灣).

Mosuowan is located at the center of Gurbantünggüt Desert (古爾班通古特沙漠), between 44°44′ and 45°01′ of north latitude, 84°52′ and 86°50′ of east longitude. Its outline is like a trumpet about 64 kilometers long flowing into the desert. Mosuowan was shaped by the water of Hutubi River (呼圖壁河) and Santun River (三屯河), but it became a dry plain because the rivers never reached this region in the first 20 years of the twentieth century.[21] In the early surveys, they had found several abandoned castles, farmlands and abandoned irrigation channels on the floodplain of Mosuowan. So it was regarded as a fertile plain with a large amount of arable land. It was very attractive for local authorities.[22] Wang Zhen proposed to cultivate it. But it was waterless. Constructing water projects for Mosuowan was mentioned in the regional plan briefly in 1952.[23]

After reorganization, the Tenth Farming Division of the People's Liberation Army was stationed in Xiaolizhuang (小李莊), a village near Dahaizi Reservoir, and prepared to explore Mosuowan in May 1953. At the same time, they repaired the reservoir and dispatched engineers to survey and draw a sketch of uncultivated land in Moauowan. In the spring of 1954, they learned clearly that there were about 73,300 ha of arable land, fertile and not too saline. So they planned to expand the old Dahaizi Reservoir to introduce more water from Manas River to Mosuowan. It was included in the draft of *Water Resources Plan of the Manas River Basin* (1954).

As mentioned above, the *Water Resources Plan of the Manas River Basin* had to be approved by the Department of Water Resources and Hydropower of the PRC before being carried out. In July 1954, Fu Zuoyi (傅作義, 1895–1974),

Minister of the Water Resources and Hydropower, and other experts came to the Manas River Basin to evaluate the plan. The experts included B. B. Voronin, a Soviet hydraulic engineer who participated in many big water project plans in China; Zhukov, another Soviet irrigational expert; and Li Fudu (李賦都, 1903–1984), Minister of Water Resources Division of Northwest Military Committee. These experts all learned about the latest ideas and technology of the water project contemporarily. They studied the plan carefully and did field research, especially at the planning sites of reservoirs. In addition, they surveyed the upper reaches of the Manas River Basin in mountain areas to consider the possibility of construction mountain reservoirs.

More than 20 days later, they began to discuss the plan with the original designers. According to many participants' memoirs, there was a heated debate about where the reservoirs should be constructed, on the plain or in the mountain. The debate became very famous later. Soviet experts, represented by B. B. Voronin, firmly opposed the construction of plain reservoirs. The main reasons for opposing plain reservoirs construction were: (1) they need much more capital; (2) dam breaking is a possible disaster; and (3) they cause the water table to rise, leading to secondary salinization in surrounding land. The last point was the core issue for Soviet experts to consider, based on the experience of Soviet Central Asia. The Soviet experts pointed out that secondary salinization soil was hard to be remedied and would make the land become barren, and the residents would be forced to emigrate. They argued that constructing reservoirs in the mountain area was the best choice for the multi-purpose of hydroelectric power generation, flood control and irrigation.[24] Opinions of Soviet experts were the latest ideas of basin-wide planning and dam construction at that time, similar to those in Western countries.[25]

Chinese hydraulic engineers and local authorities persisted to construct plain reservoirs and debated intensely with the Soviet experts. Su Zongsong and Zhang Zhonghan (張仲瀚, 1915–1980) represented Chinese hydraulic engineers and local authorities respectively. They contended that the main reasons of constructing plain reservoirs were: (1) they need less capital than mountain reservoirs; (2) the technology is fairly simple to construct; (3) to store stable amount of water in the Spring Water Overflowing Belt for irrigation; (4) they had no concrete or no machines like cranes, and not enough hydraulic engineers to construct a huge dam in the mountain, and yet no suitable site had been found for the reservoir.[26] All of the reasons were related to the core issue of how to provide enough water for the huge land reclamation in order to feed the people as soon as possible. Local water control authorities tended to concentrate on projects with rapid pay-offs.

At last, Chinese hydraulic engineers and local political authorities expressed their approval for the Soviet experts' opinions in principle. They said constructing mountain reservoirs were right for multi-purposes but it was "a beautiful prospect in the future," and could not be done at that time mainly because of construction materials shortage. They made all their efforts to persuade Soviet experts to agree with constructing plain reservoirs in the planning. Feeding people of Xinjiang troops was especially important for strengthening political control of the region. The opinions of Chinese hydraulic engineers and local political authorities represented the common alternative approach in the process of modernization. In order to achieve self-sufficiency in grain rapidly, many such simple water projects were constructed and powered yield increases in North China during the initial years of the PRC.

Soviet experts finally agreed to construct plain reservoirs in the planning, such as Moguhu (蘑菇湖), Daquangou (大泉溝), Jiahezi (夾河子), Aajihai (安集海), Liugou (柳溝), Chepaizi (車排字), Kuitun (奎屯) and so on, except Dahaizi Reservoir. It is puzzling that, in 1954, all the above approved reservoirs were just in the stage of drafting the plan while Dahaizi Reservoir had been in use for more than two years. Why did Soviet experts not only oppose firmly its expansion but also insist on abandoning it immediately? From currently available materials, we only learn the same reasons that Soviet experts had put forward for all plain reservoirs as above.[27] Why did the Soviet experts make such a decision?

Judging from the main critics of Soviet experts, remote sensing images, large-scale topographic maps and hydrogeological maps of the Manas River Basin,[28] we find Dahaizi Reservoir caused the water table to rise more easily than the others. That was the real reason why Dahaizi Reservoir was the only one being rejected. All the planning plain reservoirs were located on natural depressions, north of the Spring Water Overflowing Belt. It is a very flat plain with very fine soil and very weakly flowing underground water. To construct reservoirs on such flat plain would easily cause the water table rise. The Dahaizi depression is the worst one. The Dahaizi depression and its narrow northern region sandwiches between the Manas River and the Taxi River (塔西河), so there are many more water resources to supply underground water than the other depressions (see Figure 5.3).

To store more water in the depression could only cause a serious rise in the water table, which could easily cause serious salinization, especially given the high evaporation in such an arid region. The unique hydrological and geological conditions of Dahaizi Reservoir were the real reason that the Soviet experts opposed it.

Dahaizi Reservoir could not be expanded. Local authorities had to give up the attractive Mosuowan and disbanded the Tenth Farming Division of the People's Liberation Army. According to the memoirs, it was indeed a quite regrettable matter for Chinese water experts and local authorities.[29] What they worried most about was not secondary salinization around the plain reservoir but rather the Sino-Soviet relationships. From 1949 to 1955, China established a strategic alliance with the Soviet Union on the principle of one-sidedness. Many Soviet experts were dispatched to help Chinese construction projects. At that time, the government ordered that Soviet experts' advice had to be accepted unconditionally.[30] Wang Yizhong (王義忠), a local water expert, recalled:

> Soviet experts' advices were just commands at that time. Nobody dared to oppose them. How to carry out these proposals would be checked every year

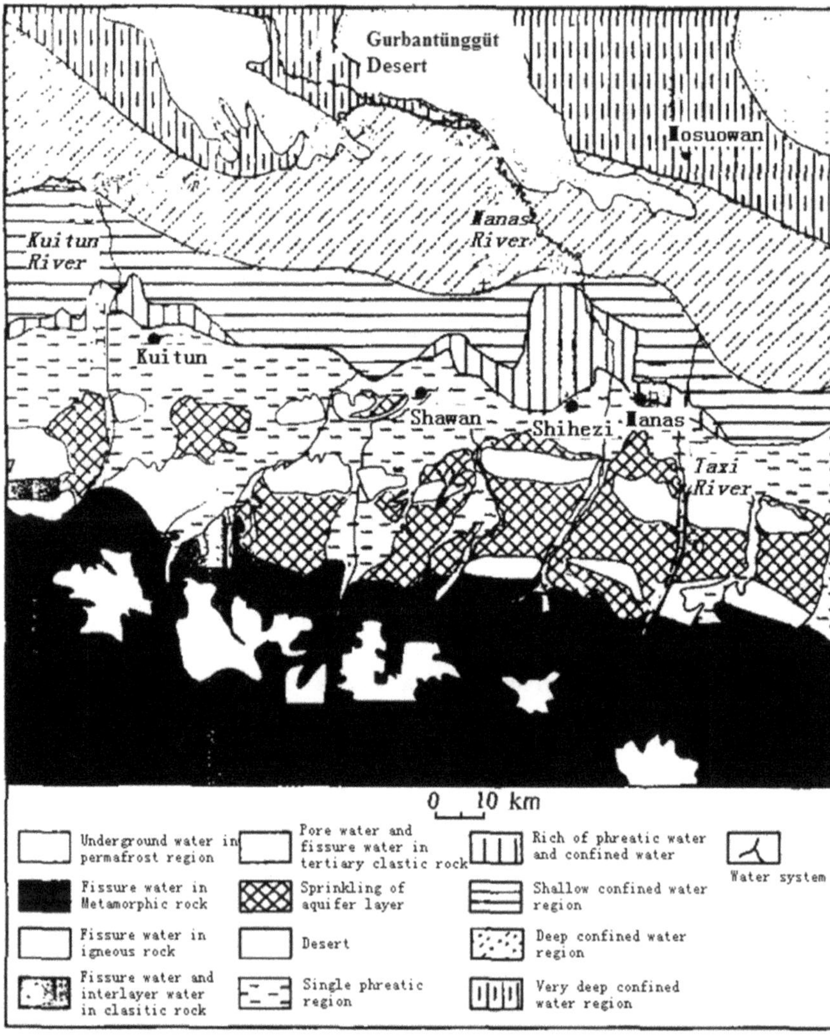

Figure 5.3 Plane map of hydrogeology for the Manas River Basin (source: Shi Xingmin et al., "Manasi he liuyu dimao yu dixiashui de guanxi (The Relationships between Manas River Basin's Geomorphology and Underground Water)," *Dili Yu Dili Xinxi Keji* [*Geography and Geo-Information Science*], No. 3 (2004): 103–120. (With permission from the author)).

> while ignoring the local actual situation completely. To put forward a different view meant opposing Soviet experts, and was regarded as undermining the Sino-Soviet relations.[31]

In fact, local water control authorities only partly executed the Soviet experts' proposals. They gave up the expansion plan but did not abandon the Dahaizi Reservoir. It was said that water table and soil were observed closely by local experts. Clashes between proponents and opponents had brought Dahaizi Reservoir into focus in Xinjiang development from 1954 to 1957, even up to 1987.[32] In 1957, they pointed out that the water table did not change in non-irrigated land, but rose rapidly in irrigated farmland in Shihutan (十戶灘), both of which were located about 15 kilometers north of Dahazi Reservoir. The water table in Shihutan rose at least from three to five meters and made the ground severely swampy not as a result of the plain reservoir, but because of irrigating water to wash salt from soil.[33]

Expanding the Dahaizi Reservoir and exploring Mosuowan

Local authorities had to give up plans of expanding Dahaizi Reservoir and exploring Mosuowan reluctantly under the influence of diplomatic policy in 1954. But later changes of domestic and international situations allowed them to surface the plans of expanding Dahaizi Reservoir and exploring Mosuowan.

On January 23, 1956, a policy of wasteland reclamation was proposed by the Political Bureau of CPC Central Committee in the "National Programme for Agricultural Development from 1956 to 1967 (Draft)."[34] Then an important editorial about "reclamation of wasteland" was issued in the *People's Daily* on March 6: reclamation of wasteland was a project of vital and lasting importance for agricultural development and would

> not only have great economic significance, but also have great political significance. There is sparse population, backward economy and culture in the northwestern and boundary provinces of China. After a planned immigration and reclamation, these areas would be prosperous. It will help to balance development level of all regions and to eliminate development imbalance in all minorities.[35]

From then on, reclamation of wasteland project was carried out in northwestern, northeastern and southwestern China.

Meanwhile, underlying changes that occurred in Sino-Soviet relationships influenced the attitude of the Chinese towards the Soviet experts' advice in practice. The symbolic event was the "Secret Speech" of Nikita Khrushchev (1894–1971) denouncing Stalin's purges on February 25, 1956, at the 20th Party Congress. Divergent opinions emerged between China and the Soviet Union. On March 24, 1956, Mao Zedong said that Khrushchev's speech against Stalin was good for China because it forced the Chinese to seek suitable policies for China's socialist construction rather than copying the Soviet Union's model completely.[36] And then the Chinese Communist Party put forward a slogan of "taking lessons from the Soviet Union." Shortly after, Xi Zhongxun (習仲勳, 1913–2002), Secretary General of the Chinese State Council, criticized thoughts of "everything depends on Soviet experts" and pointed out that 70 percent of the experts' advice was feasible while some of it was not. Such criticism made most local officials call into question the theories and experiences of the Soviet Union.[37]

Taking this chance, the Xinjiang Production and Construction Corps put the exploratory Mosuowan plan on schedule again. Local experts surveyed there, drew topographic maps, tested soil and began to draft an exploration plan in 1956.[38] According to the "reclamation wasteland" project, ten million hectares of wasteland in China must be reclaimed during the second Five-Year Plan (1958–1962). Mosuowan, which was about more than 66,700 ha wasteland, was the best choice to be developed for local authorities.[39] Lifting the fetters of obeisance to Soviet experts, the Eighth Division of Xinjiang Production and Construction Corps made the decision to expand the Dahaizi Reservoir and reclaim Mosuowan immediately. The draft of exploration plan for Mosuowan was completed in July 1957 and was approved only by the Xinjiang Production and Construction Corps on November 21, 1957. Soon, more than 10,000 workers were organized to construct the water project and thousands of people were mobilized to hand-dig channels and reclaim land in Mosuowan that winter.

Dahaizi Reservoir was expanded from May 1, 1958 to April 2, 1959. Its name was changed to Yuejin (躍進), meaning leap forward, because it was expanded at the beginning of the Great Leap Forward (1958–1961). Its capacity was 100 million cubic meters (the original was four million cubic meters) and it irrigated about 77,300 hectares (the original irrigated 533.3 hectares) of land. The expansion project was constructed in a hurry and many original designs were reduced in order to make it functional as soon as possible.

Similar to the practice of the first construction in 1952, the expansion project was also completed with many traditional methods, including human labor, low technology except for a few modern machines. However, we should notice the scale of the reservoir itself and its impacts on society, economy and environment. It was a huge reservoir with many irrigation canals, turning a large area of uncultivated land into farmland, promoting local economic development and influencing the regional ecologic system. In this sense, construction of the Dahaizi Reservoir was not a traditional but a modern event. Many water projects in China in the 1950s, even in the 1960s, were similar to the Dahaizi Reservoir.

Besides, the relationship between giving approval and the actual construction of water projects should receive more attention. From 1952 to 1956, the central government held the right of approving water projects in the Manas River Basin. All water planning was examined by hydraulic engineers carefully so that they could avoid serious problems in the future. But in 1957, the local authority, the Xinjiang Production and Construction Corps, took advantage of the delegation of approval right for water projects from the Central Government to do what they wanted to do, expanding the Dahaizi Reservoir and exploring Mosuowan.

It can be said that the expansion of the Dahaizi Reservoir, from the speedy approval to the actual construction within a short time, was just a prelude to the Great Leap Forward. Therefore, we think the delegation of approval authority for water projects to local governments during the Great Leap Forward may be a key factor for so many low-technological and poor quality water projects being constructed in this short period.[40] All of this would have a profound impact on the contemporary environmental history of China.

Environmental impacts of the Dahaizi Reservoir

The Dahaizi Reservoir plays a key role in local economic development until today, especially for Mosuowan. It has both positive and negative effects whether they were taken seriously or not by policy-makers. Benefits of the reservoir are obvious in that it has contributed significantly to increased farmland, higher production level and higher living standards for local residents, with a reliable and streamlined irrigational network. Reclamation land in Mosuowan increased rapidly from 12,000 ha in 1958 to 53,300 hectares in 1960. It was regarded as the typical case of "the Transformation of Nature (改造自然)," "simple-living and hard struggle (艱苦奮鬥)," "building socialism quickly, better and more economically (多快好省地建設社會主義)" and "March to the desert (向沙漠進軍)." The big economic achievement in Mosuowan was always emphasized and propagated.

The main negative effect on the environment is that it causes a higher water table level around the reservoir and causes secondary soil salinization which appeared immediately after the reservoir was completed. Local experts repeatedly expressed that secondary soil salinization around the Dahaizi Reservoir was not serious and could be minimized by draining canals and planting trees on the outside of the dam.[41] Besides, we notice that the local authorities realized the advantages and disadvantages of the Dahaizi Reservoir clearly before they made decisions according to the above debates. What was the core concern could be reflected vividly by the remark of a local authority, Chen Shi (陳實), the Chief of Staff of Xinjiang Production and Construction Corps. He said in 1957:

> [Dahaizi] reservoir is constructed on wasteland. No residents are nearby. Even if salinization occurs in part of the land, we can get more than 1 million *mu* (about 66,700 hectares) of land in Mosuowan. It will be a grain and cotton production base. Fishing a fish with an earthworm, does it gain more than it loses?[42]

It is very clear that local authorities held views that it was worthy to change a degraded wasteland for a huge fertile land. So they always exaggerated its benefits on society one-sidedly, but omitted the facts of the raised water table and secondary soil salinization surrounding the reservoir.

It should be discussed that one scholar had pointed out that there was no environmental evaluation before 1981 because China did not realize the issues of environmental security at that time.[43] From our case, we can say that the policy-makers were not unaware of the potential environmental damages but decided to sacrifice the environment for the economic prosperity.

In fact, the water table raises not only around the reservoir but also about 15 kilometers away, mainly in Shihutan, the farmland of the 147th Mission of the Eighth Division of Xinjiang Production and Construction Corps. It is a flood-plain where the water table was five meters deep before reclamation. Intense evaporation has caused very serious soil salinization in that place. It is a famous saline wasteland. In an early but unpublished article, researchers had pointed out that water tables rose rapidly not only in abandoned land near the Dahaizi Reservoir but also in all relatively distant farmlands of the 147th Mission. The area of land with a 2.5 meter water table increased from 2,300 hectares in 1954 to more than 9,600 hectares in 1958 (see Table 5.1).[44]

Another record can reflect the serious adverse impact of the Dahaizi Reservoir. There were 9,700 hectares of land with secondary soil salinization from 1958 to 1962; houses collapsed due to saline intrusion and the residents migrated several times.[45] Besides, the total area of farmland in the 147th Mission was 110,000 hectares in 1964 and did not increase much from then on,[46] quite different from other regions where land increased rapidly with new water technology. This shows that secondary soil salinization became the key restrictive factor for farmland increase even from the mid-1960s.

According to the classification of impacts on the surrounding area and on the distant region which the reservoir mainly services, we can say it is positive for Mosuowan region, more than 40 kilometers away from the reservoir, and negative for the surrounding area. It seemed rational for local authorities today to suggest that the surrounding land was sacrificed for the sake of agricultural prosperity in Mosuowan. We would like to emphasize that we should face directly the serious problems of environmental degradation around the reservoir as soon as possible. The adverse impacts of the reservoir should be minimized or eliminated by more careful planning and design that incorporate public involvement and input early in the process.

Table 5.1 Area of different water table in 147th Mission (ha)

Year/water table	*1954*	*1955*	*1957*	*1958*	*1959*	*1960*	*1961*	*1962*
Below 2.5 meters	2,390	3,690	6,170	9,620	11,030	13,970	11.760	18.280
2.5–4 meters	6,370	8,120	10,360	8,150	6,760	4,760	6.900	10.460
Above 4 meters	10,550	7,510	2,790	1,550	1,320	0.580	0.650	0.230

Source: Li Shugang and Zhao Huaibi, "Shihezi 147 tuan guanqu turang cisheng yanzihua xindongtai (New Dynamic Developments of Soil Salinization in Irrigation Region of 147th Mission)," *Ganhanqu Yanjiu* [*Arid Zone Research*], No. 1 (1984): 28–35.

Conclusion

Frontier consolidation was the core issue that dominated economic policies and environmental attitudes in Xinjiang throughout the 1950s. Farming around troops' garrisons was a Chinese tradition and was the most economic method for self-sufficiency in grain. With the help of Soviet experts, and more importantly with much human labor, a large amount of wasteland was reclaimed, irrigation networks were built, and cotton cultivation was successfully implemented in the Manas River Basin. All of this made it a representative of Chinese modern agriculture and was heavily promoted as a symbol of national achievement. That was the reason why the expansion of the Dahaizi Reservoir, on the Manas River Basin, attracted more attention among most water experts at that time.

In the disputes on the expansion project of the Dahaizi Reservoir, the Soviet experts emphasized the irreversible impact of plain reservoir and promoted constructing the reservoir in the mountain area for a multi-purpose in a modern way. However, Chinese policy-makers had emphasized a feasible way to build plain reservoirs because construction materials were insufficient for the mountain reservoir at that time. Whether the authorities decided to give up or conduct the expansion project at different times, they never put environmental issues into their central concerns. The policy-makers were not unaware of the environmental damage but decided to sacrifice the environment for economic prosperity. They believed it was worthy to turn a small degraded wasteland into a huge fertile farmland. So they always exaggerated its benefits on society one-sidedly, and dismissed the facts of raised water tables and secondary soil salinization surrounding the reservoir. For policy decision-makers, the most important issue was frontier security and its related Sino-Soviet relationships as well as economic development. Just as J. R. McNeill had pointed out, "Regardless of political system, policy makers at all levels from local to international responded more readily to clear and present dangers (and opportunities) than to more subtle and gradual worries about the environment."[47] Sino-Soviet relationships and boundary security were clear and present dangers while environmental effects did not get instant results or across time and space, although they had influenced the regional ecological system profoundly.

What is remarkable, however, is in that the construction of reservoirs in the 1950s, China adopted a modern approach rather than a traditional one for its profound impacts on all aspects of the region. It is the Chinese modernity in the 1950s which stands out in the story of debates on the expansion project of the Dahaizi Reservoir.

Acknowledgment

We thank Prof. David Peitz and Prof. David Bell for their thoughtful advice to revise, and thank Mr. Brian Lander and Miss Liu Wei for their careful proofreading and helpful comments and suggestions for improvement. This study was financially supported by grants from the National Science Foundation of China (41271159, 40801047) and a grant from the National Social Science Foundation of China (13FZS014).

Notes

1 J. R. McNeill, *Something New under the Sun: An Environmental History of the Twentieth-century World* (New York: W. W. Norton & Company, Inc. 2000), p. 355.
2 Nearly 700 dams were built every ten years up to the 1950s and grew rapidly after 1950s. Now there are about 45,000 large dams in the world. About half of the world dams are in China and India.
3 Judith Shapiro, *Mao's War against Nature: Politics and the Environment in Revolutionary China* (Cambridge: Cambridge University Press, 2000), pp. 21–66; David Pietz, *Engineering the State: The Huai River and Reconstruction in Nationalist China, 1927–1937* (London: Routledge, 2002), pp. 39–124; Kenneth Pomeranz, "Chinese Water Project as Economic, Environmental and Symbolic Constructions: Perspectives from Comparative History" (Paper for UNESCO UCI Groundwater conference, December 2008, Irvine, California), available at: www.groundwater-conference.uci.edu/proceedings.html, accessed on June 10, 2012; David Pietz, "Water, State, and History in the Yellow River Valley, 1949–2009" (paper presented at the First Conference of East Asia Environment History, Taipei, Taiwan, December 24–26, 2011).
4 Li Yupu 李雨普, "Exploration Water Resources in Xinjiang (新疆水利開發)," *China Water Resource* [人民水利], No. 2 (1950): 55–56.
5 "Commanders of Xinjiang Military Region of People's Liberation Army, January 12, 1950 (中國人民解放軍新疆軍區命令, 1950年1月12日)," in Xinjiang Production and Construction Corps' Compilation Committee 新疆生產建設兵團編著委員會 (ed.), *Selected Materials about Xinjiang Production and Construction Corps* [新疆生產建設兵團史料選輯], Vol. 1 (1991), pp. 36–37. Unpublished inner materials.
6 There was only one unfinished reservoir, named Hong Yan Chi, before 1949 in Xinjiang.
7 "Commanders of Xinjiang Military Region of People's Liberation Army, January 12, 1950", in *Selected Materials about Xinjiang Production and Construction Corps*, Vol. 1 (1991), pp. 36–37. Unpublished inner materials.
8 Shihezi Hydro Electricity Bureau Autonomous Region's Manas River Management Office 石河子水電局自治區瑪管處, *Water Conservancy Chronicle of Manas River Basin* [瑪納斯河流域水利志], (1992), pp. 23, 165. Unpublished inner materials.
9 Tao Zhiyue 陶峙嶽, "The History of Xinjiang Merino Entrepreneurial (新疆軍墾創業紀實)," in Chinese People's Political Consultation Conference's Committee of Cultural and Historical Data 全國政協文史資料委員會 (ed.), *Review on Exploration History of Northwestern Region* [西北開發歷史回顧], (Beijing: Zhongguo Wenshi Chubanshe, 2000), pp. 288–305.
10 Feng Shou 豐收, *Wang Zhen and Us* [王震和我們], (Urumqi: Xinjiang Renmin Chubanshe, 2008), p. 42.
11 Headquarter of the Twenty-second Corps 二十二兵團司令部, "Regional Planning of Manas River and Kuitun River, May, 1952 (瑪納斯河奎屯河區域規劃, 1952年五月," in The Compilation Committee of Chorography of the Eighth Division in Shihezi City 農八師石河子市地方誌編著委員會 (ed.), *Chorography of the Eighth Division of Xinjiang Production and Construction Corps in Shihezi City* [農八師墾區石河子市志], (Urumqi: Xinjiang Renmin Chubanshe, 1994), pp. 826–828.
12 "Great Plan for the Transformation of Nature: Report of the Council of Minister and the Central Committee of the Soviet Union Communist Party (蘇聯部長會議和聯共中央報告：改造大自然的龐大計劃)," *Compilation of Selected News and Repots* [新聞類編], No. 1683 (July 17, 1949).

13 Propaganda Newsgroup of Xinjiang Military District 新疆軍區政治宣傳新聞組, "Our Defence Forces Are Also Production Military—The Great Achievements of People's Liberation Army Stationed in Xinjiang on the Economic Front During Three Years (我們的國防軍也是生產軍—駐新疆的人民解放軍三年來在經濟戰線上的偉大成就)," *People's Daily* [人民日報], A02, October 22, 1952.

14 Xinjiang Uygur Autonomous Region's Office of Water Resources, First Team of Investigation and Design 新疆維吾爾自治區水利廳第一勘測設計隊, *The Survey Report of Moguhu Reservoir and Others in Manas River Basin, January 1953* [瑪納斯和流域蘑菇湖水庫及其他水庫查勘報告 (1953年1月)], an archive preserved in Xinjiang Institute of Ecology and Geography, Chinese Academy of Sciences (中國科學院新疆生態與地理研究所).

15 The area of farmland was more than 20,000 hectares before 1949. See *Water Conservancy Chronicle of Manas River Basin*, (1992), p. 165. Unpublished inner materials.

16 Zhu Haiguang 朱海光 et al., "The Comments On the Irrigation of Xinjiang (對於新疆省農田水利工作的意見)," *China Water Resource* [人民水利], No. 2 (1953): 63–64.

17 Cheng Zhaoxuan 程照軒, "Main Experiences for Cotton Production in Xinjiang (新疆棉花大面積豐產的主要經驗)," *People's Daily* [人民日報], A02, December 13, 1995; Xinjiang Uygur Autonomous Region's Planning Committee 新疆維吾爾自治區計劃委員會 (ed.), *Xinjiang Uygur Autonomous Region's National Economic Data (1949–1957)* [新疆維吾爾自治區國民經濟資料 (1949–1957)], July 1959, p. 95. Unpublished inner materials.

18 "A Large Area of Cotton in Manas River Basin Got Good Harvest (新疆瑪納斯河流域棉花獲得大面積豐產)," *People's Daily* [人民日報], A02, December 5, 1953.

19 Reports on *People's Daily* were: Yang Shufan 楊樹藩, "Analyze the Cotton Yield Potential Through the Increased Production in 1953 (從一九五三年棉花增產的範例中看挖掘棉花增產潛力的途徑)," A02, March 26, 1954; Tupian baodao 圖片報導, "A Large Area of Cotton in Manas River Basin Got Good Harvest (瑪納斯河流棉花大面積豐產)," January 12, 1954; Yang Xiandong 楊顯東, "Development of Our Cotton Production 我們棉花生產的發展)," September 20, 1954; Shen Shi 沈石, "In Manas River Basin (在瑪納斯河流域)," August 9, 1955; Jin Tong 金同, "Pioneers for Happy Life (幸福生活的開拓者)," September 19, 1955; Han Jiang 寒江, "To Produce More Grain and Cotton in North and South of Tianshan Mountains (讓天山南北生產更多的糧食和棉花)," A06, October 1, 1955; Shen Shi 沈石, "How to Get Good Harvest (豐收是怎樣得來的)," October 27, 1955; Xinhuashe, 新華社, "Promote the Cotton Harvest Experience of Xinjiang Army in the Countrywide (全國將推廣新疆部隊棉花大面積豐產經驗)," A01, November 4, 1955; Liu Kairu 劉開儒, Wu Chunzhan 武純展 and Tan Xin 譚忻, "Good Cotton Harvest of Eighty Thousand *mu* land by Xinjiang Production Troops (新疆生產部隊八萬多畝棉田豐收)," A02, November 6, 1955; Cheng Zhaoxuan 程照軒, "Main Experiences for Cotton Production in Xinjiang (新疆棉花大面積豐產的主要經驗)," A02, December 13, 1955, and so on.

20 Su Zongsong was born in 1910. He graduated from Zhejiang University in 1934 and served in the Water Resources Department of National Economic Council from 1934 to 1940. He had visited Egypt to learn the irrigation technology in dry area from 1937 to 1939. During this period, he saw several big water projects being built on the Nile. From 1950 to 1956, he worked in the Water Resources and Hydropower of PRC. See Li Chengxiu 李成秀, "Su Zongsong 粟宗嵩," in China Association for Science and Technology 中國科學技術學會 (ed.), *The Biography of Chinese Scientific and Technical Experts Agronomy Volume Comprehensive, Volume 1* [中國科學技術專家傳略· 農學編· 綜合卷1], (Beijing: Zhongguo Nongye Kexue Jishu Chubanshe, 1996), pp. 408–421.

21 Zhang Li 張莉 et al., "Changes of Santun River and Hutubi Rive in the Past 300

Years, Xinjiang, China (近300年來新疆三屯河與呼圖壁河水系變遷研究)," *Beijin Acta Scientiarum Naturalium Universitatis Pekinensis* [北京大學學報 (自然科學版)], Vol. 40, No. 5 (November 2004): 957–970.

22 Shi Zhen 石震, "History of Mosuowan Exploration (莫索灣創業史話)," in *Sons of Tianshan: Collections of Essays and Reportages* [散文報告文學集—天山之子], (Beijing: Jiefangjun Chubanshe, 1984), pp. 485–499. It was written on May 34, 1964.

23 Headquarter of the Twenty-second Corps, "Regional Planning of Manas River and Kuitun River, May, 1952," in *Chorography of the Eighth Division of Xinjiang Production and Construction Corps in Shihezi City*, pp. 826–828.

24 Wang Yizhong 王義忠, "The Initial Development of the Manas River Basin (瑪納斯河流域的初期開發)," in Gong Yuzhi 龔育之 (ed.), *Review on Exploration History of Northwestern Region* [西北開發歷史回顧], (Beijing: Zhonguo Wenshi Chubanshe, 2000), pp. 306–314; Wang Yizhong 王義忠, "Recalling the Exploration and Water Construction in Shihezi Farming Region (憶石河子墾區的開發和水利建設)," in Li Kaiquan 李開全 (ed.), *Selected Materials about Xinjiang Production and Construction Corps*, Vol. 15, (Urumqi: Xinjiang Renmin Chubanshe, 2005), pp. 228–242; Li Chengxiu, "Su Zongsong," p. 413; Xiong Kunjing 熊坤靜, "General Zhang Zhonghan's Emotional Ties Toward Farming Region (張仲瀚將軍情系屯墾半生緣)," *Materials about Culture and History* [文史精華], No. 8 (2007): 25–31.

25 Plannings of micro rivers have been developed over the last three decades for multipurpose. See International Commission on Irrigation and Drainage, "Role of Dams for Irrigation, Drainage and Flood Control," ICID position paper, available at www.icid.org/dam_pdf.pdf, accessed on September 10, 2013; Lubiao Zhang, "Social Impacts of Large Dams: The China Case", A working paper of the World Commission on Dams, 2000, available at: http://citeseerx.ist.psu.edu/viewdoc/Download?doi=10.1.1.117.7440&rep=rep1&type=pdf, accessed on September 10, 2013; B. B. Voronin 沃洛寧, "The Method to Plan a Comprehensive Utilization of Water and Land Resources of River Basin (水利和土地資源綜合利用的流域規劃及其編制辦法)," *Chinese Science Bulletin* [科學通報] No. 5 (1954): 26–34.

26 Wang Yizhong, "The Initial Development of the Manas River Basin," pp. 306–314; Wang Yizhong, "Recalling the Exploration and Water Construction in Shihezi Farming Region," pp. 228–242; Li Chengxiu, "Su Zongsong," p. 413; Xiong Kunjing, "General Zhang Zhonghan's Emotional Ties Toward Farming Region)," *Materials about Culture and History* [文史精華], No. 8 (2007): 25–31.

27 We tried to access the files to learn the special objection reasons for Dahaizi in local archives in the summer of 2011. But we were told that these files are still un-decrypted because they relate to the Sino-Soviet relations.

28 Shi Xingmin 史興民 et al., "The Relationships Between Manas River Basin's Geomorphology and Underground Water (瑪納斯河流域地貌與地下水的關係)," *Geography and Geo-Information Science* [地理與地理信息系統] No. 3 (2004): 103–120.

29 Wang Yizhong, "The Initial Development of the Manas River Basin," pp. 306–314; Wang Yizhong, "Recalling the Exploration and Water Construction in Shihezi Farming Region," pp. 228–242; Li Chengxiu, "Su Zongsong," p. 413; Xiong Kunjing, "General Zhang Zhonghan's Emotional Ties Toward Farming Region," pp. 25–31.

30 Shen Zhihua 沈志華, "Historical Investigation on Soviet Experts—Basic Conditions and Policy Changes (對在華蘇聯專家問題的歷史考察—基本狀況及政策變化)," *Contemporary China History Studies* [當代中國史研究], No. 1 (2002): 38–44.

31 Feng Shou, *Wang Zhen and Us*, p. 42.

32 Xinjiang Uygur Autonomous Region's Association of Water Resources 新疆維吾爾自治區水利學會, "Discussion on Plain Reservoir in Xinjiang (關於新疆平原水庫的討論)," *Irrigation and Drainage* [灌溉排水], No. 3 (1984): 30–35. This paper was written by Wang Heting 王鶴亭.

33 Xinjiang Uygur Autonomous Region's Office of Water Resources 新疆維吾爾自治區水利廳, "Introduction to the Water Resources Plan of Manas River Basin (瑪納斯

河流域規劃介紹)," in the Ministry of Water Resources' Investigation and Design Bureau 水利部勘測設計局 (ed.), *Intoduction to Ordinary Experience: Compilation of Selected Documents to Exchange from the Conference on Medium and Small River Planning* [全國中小流域規劃會議經驗交流文件彙編之一: 一般經驗介紹], (Beijing: Shuili Dianli Chubanshe, 1958), pp. 65–75.

34 Central Politburo of the Communist Party of China 中共中央政治局, "National Agriculture Development Program From 1956 to 1967 (Draft), January 23, 1956 (一九五六到一九六七年全國農業發展綱要(草案), 1956年1月23日)," in CCCPC Party Literature Research Office 中共中央文獻研究室 (ed.), *Compilation of Selected Important Documents since 1949* [建國以來重要文獻選編], Vol. 8, (Beijing: Zhongyang Wenxian Chubanshe, 1993), pp. 46–60.

35 "Editorial on Reclamation Wasteland (開墾荒地社論)," *People's Daily* [人民日報], March 6, 1956.

36 Wu Lengxi 吳冷西, *Ten Years Debating—Memoirs about Sino-Soviet Relationship, 1956–1966* [十年論戰—1956–1966中蘇關係回憶錄], (Beijing: Zhongyang Wenxian Chunbanshe, 1999), pp. 14–15.

37 Shen Zhihua 沈志華, "Historical Survey on Soviet Experts' Role and Influence in China—Basing on Archival Documents and Oral Historical materials in China and Russia (對在華蘇聯專家問題的歷史考察: 作用和影響—根據中俄雙方的歷史檔案文獻和口述史料)," *Journal of Chinese Communist Party History Studies* [中共黨史研究], No. 2 (2002): 38–44.

38 Xinjiang Uygur Autonomous Region's Chorography Compilation Committee and Xinjiang Production and Construction Corps' Chronical Compilation Committee 新疆維吾爾自治區地方誌編著委員會、《新疆通志· 新疆生產建設兵團志》編著委員會 (eds.), *Annals of Xinjiang* [新疆通志], Vol. 37, *Chronicles of Production and Construction Corps* [生產建設兵團志], (Urumqi: Xinjiang Renmin Chubanshe,1998), p. 103.

39 Han Zuobin 韓作斌 (ed.), *Chronicles of 148th Mission of Eighth Division of Xinjiang Production and Construction Corps* [新疆生產建設兵團農八師一四八團團場志], (Urumqi: Xinjiang Renmin Chubanshe, 1997), p. 64.

40 There are another two materials that can support our opinion about the relations between approval authority for water projects delegated and many low-technological and poor quality water projects. This issue is worth further study. See: Compilation Committee of Series of Modern Guangxi [當代廣西] 叢書編委會, *Water Projects Construction in Modern Guangxi* [當代廣西水利建設], (Beijing: Dangdai Zhongguo Chubanshe, 1997), p. 24; Li Chuanshou 李傳授 (ed.), *Annals of Dazu County* [大足縣誌], (Beijing: Fangzhi Chubanshe, 1996), p. 688.

41 Xinjiang Uygur Zizhiqu Shuili Xuehui, Wang Heting (writer), "Discussion on Plain Reservoir in Xinjiang (關於新疆平原水庫的討論)," *Irrigation and Drainage* [灌溉排水], No. 3 (1984): 30–35; Wang Yizhong, "The Initial Development of the Manas River Basin," pp. 306–314.

42 Shi Zhen, "History of Mosuowan Exploration," pp. 485–499.

43 Lubiao Zhang, "Social Impacts of Large Dams: The China Case," a working paper of the World Commission on Dams, 2000, available at: http://citeseerx.ist.psu.edu/viewdoc/download?doi=10.1.1.117.7440&rep=rep1&typq=pdf, accessed on September 10, 2013.

44 Li Shugang 李述剛 and Zhao Huaibi 趙懷璧, "New Dynamic Developments of Soil Salinization in Irrigation Region of 147th Mission (石河子147團灌區土壤次生鹽漬化新動態)," *Arid Zone Research* [乾旱區研究], No. 1 (1984): 28–35.

45 Wu Xiaokui 吳曉槐 (ed.), *Chronicles of 147th Mission of Eighth Division of Xinjiang Production and Construction Corps* [新疆生產建設兵團農八師一四七團場志], (Urumqi: Xinjiang Renmin Chubanshe, 1999), pp. 138–139.

46 Ibid.

47 J. R. McNeill, *Something New under the Sun*, p. 355.

Bibliography

Central Politburo of the Communist Party of China (Zhongguo Zhongyang Zhengzhiju 中共中央政治局). "Yijiuwuliu nian dao yijiuliuqi nian quanguo nongye fazhan gangyao (cao an) (Jan. 23, 1956) 一九五六到一九六七年全國農業發展綱要(草案), 1956年1月23日 (National Agriculture Development Program From 1956 to 1967 (Draft), Jan. 23, 1956)," in CCCPC Party Literature Research Office (Zhonggong Zhongyang Wenxian Yanjiushi 中共中央文獻研究室) (ed.). *Jianguo Yilai Zhongyao Wenxian Xuanbian* 建國以來重要文獻選編 [*Compilation of Selected Important Documents since 1949*], Vol. 8. Beijing: Zhongyang Wenxian Chubanshe, 1993, pp. 46–60.

Chen Weiming 程維民, Zhou Chenhu 周成虎, Liu Haijiang 劉海江, Zhang Yang 張暘, Jiang Yan 蔣豔, Zhang Yichi 張一馳 and Yao Yonghui 姚永慧. "Manas he liuyu wushi nian luzhou kuozhan ji shengtai huanjing yanbian yanjiu 瑪納斯河流域50年綠洲擴張及生態環境演變研究 (Oasis Expansion and Eco-environmental Evolution of Manas River Basin in the Past 50 Years)," *Zhongguo Kexue D: Diqiu Kexue* 中國科學 D: 地球科學 [*Science in China Series D: Earth Sciences*], Vol. 35, No. 11 (2005): 1074–1086.

Compilation Committee of Series of Modern Guangxi (*Dangdai Guangxi Congshu Bianweihui* [當代廣西] 叢書編委會). *Dangdai Guangxi Shuili Jianshe* 當代廣西水利建設 [*Water Projects Construction in Modern Guangxi*]. Beijing: Dangdai Zhongguo Chubanshe, 1997.

Feng Shou 豐收. *Wang Zhen he Wo Men* 王震和我們 [*Wang Zhen and Us*]. Urumqi: Xinjiang Renmin Chubanshe, 2008.

Han Zuobin 韓作斌 (ed.). *Xinjiang shengchan jianshe bingtuan nongbashi yisibatuan tuanchang zhi* 新疆生產建設兵團農八師一四八團團場志 [*Chronicles of 148th Mission of Eighth Division of Xinjiang Production and Construction Corps*]. Urumqi: Xinjiang Renmin Chubanshe, 1997.

Headquarter of the Twenty-second Corps (Ershier Bing Tuan Silingbu 二十二兵團司令部). "Manasi he Kuitun he quyu guihua, May, 1952 瑪納斯河奎屯河區域規劃，1952年5月 (Regional Planning of Manas River and Kuitun River)," in The Compilation Committee of Chorography of the Eighth Division in Shihezi City (Nongbashi Shihezi Shi Difangzhi Bianzuan Weiyuanhui 農八師石河子市地方誌編著委員會) (ed.), *Nongbashi Kenqu Shihezi Shizhi* 農八師墾區石河子市志 [*Chorography of the Eighth Division of Xinjiang Production and Construction Corps in Shihezi City*]. Urumqi: Xinjiang Renmin Chubanshe, 1994, pp. 826–828.

International Commission on Irrigation and Drainage, "Role of Dams for Irrigation, Drainage and Flood Control," ICID position paper. Available at: www.icid.org/dam_pdf.pdf, accessed on September 10, 2013.

Li Chengxiu 李成秀. "Su Zongsong 粟宗嵩," in China Association for Science and Technology (Zhongguo Kexue Jishu Xiehui 中國科學技術學會) (ed.). *Zhongguo kexue jishu zhuanjia zhuanlue·nongye bian·zonghe, juan 1*中國科學技術專家傳略·農學編·綜合卷1 [*The Biography of Chinese Scientific and Technical Experts·Agronomy Volume·Comprehensive, Volume 1*]. Beijing: Zhongguo Nongye Kexue Jishu Chubanshe, 1996.

Li Chuanshou 李傳授 (ed.). *Dazu Xianzhi* 大足縣誌 [*Annals of Dazu County*]. Beijing: Fangzhi Chubanshe, 1996.

Li Shugang 李述剛 and Zhao Huaibi 趙懷璧. "Shihezi 147 tuan guanqu turang cisheng yanzihua xindongtai 石河子147團灌區土壤次生鹽漬化新動態 (New Dynamic Developments of Soil Salinization in Irrigation Region of 147th Mission)," *Ganhanqu Yanjiu* 乾旱區研究 [*Arid Zone Research*], No. 1 (1984): 28–35.

Li Yupu 李雨普. "Xinjiang sheng shuili gaikuang 新疆水利開發 (Exploration Water Resources in Xinjiang)," *Renmin Shuili* 人民水利 [*China Water Resource*], No. 2 (1950): 55–56.

Lubiao Zhang. "Social Impacts of Large Dams: The China Case," a working paper of the World Commission on Dams, 2000. Available at: http://citeseerx.ist.psu.edu/viewdoc/download?doi=10.1.1.117.7440&rep=rep1&type=pdf, accessed on September 10, 2013.

McNeill, J. R. *Something New under the Sun: An Environmental History of the Twentieth-century World.* New York: W. W. Norton & Company, Inc. 2000.

Pietz, David. *Engineering the State: The Huai River and Reconstruction in Nationalist China, 1927–1937*. London: Routledge, 2002.

Pietz, David. "Water, State, and History in the Yellow River Valley, 1949–2009," paper presented at the First Conference of East Asia Environment History, Taipei, Taiwan, December 24–26, 2011.

Pomeranz, Kenneth. "Chinese Water Project as Economic, Environmental and Symbolic Constructions: Perspectives from Comparative History." Paper presented at UNESCO UCI Groundwater conference, December 2008. Irvine, California. Available at: www.groundwater-conference.uci.edu/proceedings.html, accessed on June 10, 2012.

Shapiro, Judith. *Mao's War against Nature: Politics and the Environment in Revolutionary China.* Cambridge: Cambridge University Press, 2000.

Shen Zhihua 沈志華. "Dui zaihua sulian zhuanjia wenti de lishi kaocha—jiben zhuangkuang ji zhengce bianha 對在華蘇聯專家問題的歷史考察—基本狀況及政策變化 (Historical Investigation on Soviet Experts—Basic Conditions and Policy Changes)," *Dangdai Zhongguo Shi Yanjiu* 當代中國史研究 [*Contemporary China History Studies*], No. 1 (2002): 38–44.

Shen Zhihua 沈志華. "Dui zaihua sulian zhuanjia wenti de lishi kaocha: zuoyong he yingxiang—genju zhonge shuangfang de dangan wenxian he koushu shiliao 對在華蘇聯專家問題的歷史考察：作用和影響—根據中俄雙方的歷史檔案文獻和口述史料 (Historical Survey on Soviet Experts' Role and Influence in China—Basing on Archival Documents and Oral Historical materials in China and Russia)," *Zhonggong Dangshi Yanjiu* [*Journal of Chinese Communist Party History Studies*], No. 2 (2002): 38–44.

Shi Xingmin 史興民, Yang Jingchun 楊景春, Li Youli 李有利 and Nan Feng 南峰. "Manasi he liuyu dimao yu dixiashui de guanxi 瑪納斯河流域地貌與地下水的關係 (The Relationships between Manas River Basin's Geomorphology and Underground Water)," *Dili Yu Dili Xinxi Keji* 地理與地理信息系統 [*Geography and Geo-Information Science*], No. 3 (2004): 103–120.

Shi Zhen 石震. "Mosuowan chuangye shihua 莫索灣創業史話 (History of Mosuowan exploration)," in *Sanwen Baogao Wenxue Ji-Tianshan Zhi Zi* 散文報告文學集——天山之子 [*Sons of Tianshan: Collections of Essays and Reportages*]. Beijing: Jiefangjun Chubanshe, 1984. pp. 485–499.

Shihezi Hydro Electricity Bureau Autonomous Region's Manas River Management Office (Shihezi shuidianju zizhiqu maguanchu 石河子水電局自治區瑪管處). *Manasi He Liuyu Shuili Zhi* 瑪納斯河流域水利志 [*Water Conservancy Chronicle of Manas River Basin*], 1992. Unpublished inner materials.

"Sulian buzhang huiyi he liangong zhongyang baogao: gaizao daziran de pangda jihua 蘇聯部長會議和聯共中央報告：改造大自然的龐大計劃 (Great Plan for the Transformation of Nature: report of the Council of Minister and the Central Committee of the Soviet Union Communist Party)," *Xinwen leibian* 新聞類編[*Compilation of Selected News and Repots*], No. 1683 (July 17, 1949).

Tao Zhiyue 陶峙嶽. "Xinjiang junken chuangye jishi 新疆軍墾創業紀實 (The History of Xinjiang Merino Entrepreneurial)," in Chinese People's Political Consultation Conference's Committee of Cultural and Historical Data (Quanguo Zhengxie Wenshi Ziliao Weiyuanhui 全國政協文史資料委員會) ed., *Xibbei Kaifa Lishi Huigu* 西北開發歷史回顧 [*Review on Exploration History of Northwestern Region*]. Beijing: Zhongguo Wenshi Chubanshe, 2000, pp. 288–305.

Voronin, B. B. 沃洛寧. "Shuili he tudi ziyuan zonghe liyong de liuyu guihua jiqi bianzhi banfa 水利和土地資源綜合利用的流域規劃及其編制辦法 (The Method to Plan a Comprehensive Utilization of Water and Land Resources of River Basin)," *Kexue Tongbao* 科學通報 [*Chinese Science Bulletin*], No. 5 (1954): 26–34.

Wang Yizhong 王義忠. "Manasi he liuyu de chuqi kaifa 瑪納斯河流域的初期開發 (The Initial Development of the Manas River Basin)," in Gong Yuzhi 龔育之 (ed.), *Xibei Kaifa Lishi Huigu* 西北開發歷史回顧 [*Review on Exploration History of Northwestern Region*]. Beijing: Zhonguo Wenshi Chubanshe, 2000, pp. 306–314.

Wang Yizhong 王義忠. "Yi Shihezi kenqu de kaifa he shuili jianshe 憶石河子墾區的開發和水利建設 (Recalling the Exploration and Water Construction in Shihezi Farming Region)," in Li Kaiquan 李開全 (ed.). *Xinjiang shengchan jianshe bingtuan shiliao xuanji* 新疆生產建設兵團史料選輯 [*Selected Materials about Xinjiang Production and Construction Corps*], Vol. 15. Urumqi: Xinjiang Renmin Chubanshe, 2005, pp. 228–242.

Wu Lengxi 吳冷西. *Shinian lunzhan—1956–1966 zhongsu guanxi huiyilu* 十年論戰—1956–966中蘇關係回憶錄 [*Ten Years Debating—Memoirs about Sino-Soviet Relationship, 1956–1966*]. Beijing: Zhongyang Wenxian Chunbanshe, 1999.

Wu Xiaokui 吳曉槐 (ed.). *Xinjiang shengchan jianshe bingtuan nongbashi yisiqi tuanzhi* 新疆生產建設兵團農八師一四七團場志 (*Chronicles of 147th Mission of Eighth Division of Xinjiang Production and Construction Corps*). Urumqi: Xinjiang Renmin Chubanshe, 1999.

Xinjiang Uygur Autonomous Region's Association of Water Resources (Zizhiqu Shuili Xuehui 新疆維吾爾自治區水利學會). "Guanyu Xinjiang pingyuan shuiku de taolun 關於新疆平原水庫的討論 (Discussion on Plain Reservoir in Xinjiang)," *Guangai Paishui* 灌溉排水 [*Irrigation and Drainage*], No. 3 (1984): 30–35.

Xinjiang Uygur Autonomous Region's Chorography Compilation Committee and Xinjiang Production and Construction Corps' Chronical Compilation Committee (Xinjiang Uygur Zizhiqu Difangzhi Bianzuan Weiyuanhui and Xinjiang Tongzhi zhi Shengchan Jianshe Bingtuan Zhi Bianzuan Weiyuanhui 新疆維吾爾自治區地方誌編著委員會、[新疆通志·新疆生產建設兵團志] 編著委員會) (eds.), *Xinjiang tongzhi* 新疆通志 [*Annals of Xinjiang*], Vol. 37, *Shengchan jianshe bingtuan zhi* 生產建設兵團志 [*Chronicles of Production and Construction Corps*]. Urumqi: Xinjiang Renmin Chubanshe, 1998.

Xinjiang Uygur Autonomous Region's Office of Water Resources (Xinjiang Uygur Zizhiqu Shuili Ting 新疆維吾爾自治區水利廳). "Manasi he liuyu guihua jieshao 瑪納斯河流域規劃介紹 (Introduction to the Water Resources Plan of Manas River Basin)," in The Ministry of Water Resources' Investigation and Design Bureau (Shuilibu Kance Shejiju 水利部勘測設計局) (ed.), *Quanguo Zhongxiao Liuyu Guihua Huiyi Jingyan Jiaoliu Wenjian Huibian Zhiyi—Yiban Guihua Jingyan Jieshao* 全國中小流域規劃會議經驗交流文件彙編之一·一般經驗介紹 [*Intoduction to Ordinary Experience: Compilation of Selected Documents to Exchange from the Conference on Medium and Small River Planning*]. Beijing: Shuili Dianli Chubanshe, 1958. pp. 65–75. Unpublished inner materials.

Xinjiang Uygur Autonomous Region's Office of Water Resources, First Team of Investigation and Design (Zizhiqu Shuiliting Diyi Kance Shejidui 新疆維吾爾自治區水利廳第一勘測設計隊). *Manasi he liuyu Moguhu shuiku ji qita shuiku chakan baogao* 瑪納斯和流域蘑菇湖水庫及其他水庫查勘報告 (1953年1月) [*The Survey Report of Moguhu Reservoir and Others in Manas River Basin, Jan. 1953*], an archive preserved in Zhongguo Kexueyuan Xinjiang Shengtai Yu Dili yanjiusuo 中國科學院新疆生態與地理研究所.

Xinjiang Uygur Autonomous Region's Planning Committee (Xinjiang Uygur Zizhiqu Jihua Weiyuanhui 新疆維吾爾自治區計劃委員會) ed. *Xinjiang Uygur Zizhiqu guomin jingji ziliao (1949–1957)* 新疆維吾爾自治區國民經濟資料 (1949–1957) [*Xinjiang Uygur Autonomous Region's National Economic Data (1949–1957)*]. July 1959. Unpublished inner materials.

Xiong Kunjing 熊坤靜. "Zhang Zhonghan jiangjun qingxi tunken banshengyuan 張仲瀚將軍情系屯墾半生緣 (General Zhang Zhonghan's Emotional Ties toward Farming Region)," *Wenshi Jinghua* 文史精華 [*Materials about Culture and History*], No. 8 (2007): 25–31.

Yuan Yongsong 袁永松 (ed.). *Xinjiang shengchan jianshe bingtuan nongbashi yiwuling tuanchang zhi* 新疆生產建設兵團農八師一五零團場志 [*Chronicles of 150th Mission of Eighth Division of Xinjiang Production and Construction Corps*]. Urumqi: Xinjiang Renmin Chubanshe, 2000.

Zhang Li 張莉, Han Guanhui 韓光輝 and Yan Dongkai 閻東凱. "Jin 300 nian lai Xinjian Santun he yu Hutubi he shuixi bianqian yanjiu 近300年來新疆三屯河與呼圖壁河水系變遷研究 (Changes of Santun River and Hutubi Rive in the Past 300 Years, Xinjiang, China)," *Beijin Daxue Xuebao (Ziran Kexue Ban)* 北京大學學報(自然科學版) [*Acta Scientiarum Naturalium Universitatis Pekinensis*], Vol. 40, No. 5 (Nov., 2004): 957–970.

"Zhongguo renming jiefangjun Xinjiang junqu minglin, January 12, 1950 中國人民解放軍新疆軍區命令，1950年1月12日 (Commanders of Xinjiang Military Region of People's Liberation Army," in Xinjiang Production and Construction Corps' Compilation Committee (Xinjiang Shengchan Jianshe Bintuan Bianzhu Weiyuanhui 新疆生產建設兵團編著委員會) (ed.), *Xinjiang Shengchan Jianshe Bintuan Shiliao Xuanji* 新疆生產建設兵團史料選輯 [*Selected Materials about Xinjiang Production and Construction Corps*], Vol. 1 (1991), pp. 36–37. Unpublished inner materials.

Zhu Haiguang 朱海光, Huang Wenji 黃文驥 and Wang Jiaheng 汪嘉衡. "Duiyu Xinjiang Sheng nongtian shuili gongzuo de yijian 對於新疆省農田水利工作的意見 (The Comments on the Irrigation of Xinjiang)," *Renmin Shuili* 人民水利 [*China Water Resource*], No. 2 (1953): 63–64.

6 Going green from industrial dream?

Community-NGOs cooperation among indigenous communities in Hualien, Taiwan

Chun-Chieh Chi

Introduction

The development path of Hualien (花蓮) County in eastern Taiwan over the past 30 years continued to be presented through two conflicting images. On the one hand, Hualien is a remote and backward "back mountain (後山)" of Taiwan that awaits the wheel of industrial development. On the other hand, it represents "the last pristine landscape" of the over-developed Taiwan. Located on the east coast of Taiwan, Hualien County is the largest county in Taiwan, covering one-eighth of Taiwan's total land area. The county is characterized by mountainous topography, with only 9 percent of its total land area below 100 meters of elevation; this is part of the reason why the population density in Hualien is relatively low compared to most other counties in Taiwan. With the Central Mountain Range to its west and the Coastal Range to its east, the main agricultural production area in Hualien is located in Huadong Rift Valley. A primarily resource production-based county, Hualien has been declining in population since the 1970s, when Taiwan's industrial development proceeded in its full strength.

In the late 1980s, faced with the declining production of primary resource in Hualien area, both central and local governments began to seek industrial growth by following the path of Taiwan's western plains. The "relocation of industries eastward" master plan was the product of such policy, as the central government tried to encourage and the local government tried to attract industries to invest in Hualien County. Unfortunately (or fortunately, for nature's sake), the "relocation" master plan was only successful in attracting cement production companies to invest in Taiwan's designated "cement production specialized zone," and the expansion of an existing cement production factory in populated Hualien City.

With the failed attempt to attract industries, Hualien County government began to promote tourism development in the county in the mid-1990s. In the 1998 County Policy Whitepapers, the goal was clearly stated as: "To construct the county into the 'Tourism Center of Asia-Pacific' by internationalizing the county with knowledge-based theme, and centered on smokestack-free industry." In 2001, the new county governor established a new Bureau of Tourism to administer tourism affairs, and to engage in tourism alliance building with two neighboring counties, Yilan (宜蘭) and Taitong (臺東). The county seems to have turned away from its industrial dream, embracing tourism by making better use of its natural beauty and cultural diversity.

Similarly, the central government initiated a national development project, "Challenge 2008—National Development Plan (2002–2007)" in which "doubling tourist arrivals" was one of the major goals. To achieve this goal, Hualien and Taitong area would become a primary tourist destination, with their pristine natural beauty such as the Costal Range, East Rift Valley, and Taroko National Park ready to attract tourists. As such, both the county and central governments worked hand in hand to promote tourism development in the eastern part of Taiwan.

Long before the tourism development policy, Hualien County was already attracting both domestic and international tourists due to its natural as well as cultural characteristics. However, tourist expenditures went mainly to big enterprises such as travel agencies, large hotels, bus companies, and theme parks, rather than local communities. This was mainly because prior to 2000, tourist activities in Hualien performed on a "mass tourism" scale with escorted tour groups leading the market. It is only in recent years, with the new trend of "ecotourism" and "cultural tourism," that more people started to engage in small-scale tourist activities. As such, many communities, previously not attractive or not easily accessible to mass tourists, have begun to engage in community tourism development. This is particularly true with many indigenous communities, many of which possess natural beauty and traditional culture, making them appealing new destinations for people interested in ecological and cultural tourism.

This chapter will discuss indigenous communities' attempts to make their communities more ecologically friendly, revive their traditional culture, and attract tourism. It will also examine in detail how local and national non-governmental organizations (NGOs) interact with indigenous communities to make their development projects more sustainable.

This research began in August 2008 and continued through July 2012. Major research methods employed include literature review, participant observation, and field research. The author and his assistants visited the research sites in the first stage of the study to investigate the natural environment, social environment, major NGOs, and leading actors, including traditional leaders, administrative officials, religious and cultural leaders, as well as operators of ecotourism. Interviews were then conducted with the aforementioned organizations and community leaders; a total of 18 individual interviews were conducted.

In addition to fieldwork and personal interviews, this study was combined with two studies in the same joint project funded by the National Science Council. Researchers and research assistants from all three subprojects visited each of the research sites together; each visit was followed by group discussions involving researchers, NGO representatives, as well as community representatives from the research sites. A third focus group discussion was held in the author's university to discuss common issues related to all research sites, such as administrative methods, tourist control, and negotiations with county government.

Biodiversity and cultural diversity

In the 1992 Earth Summit, 160 countries signed the "Convention on Biological Diversity" treaty, which became effective on December 29, 1993, and hence made biodiversity an important international concern.[1] While the main focus of biodiversity rests on three levels of the ecological system—genetic diversity, species diversity, and ecosystem diversity[2]—more and more researchers and NGOs have started to look into the linkage between "biodiversity" and "cultural diversity," and the role of the world's indigenous peoples in this relationship.

According to Durning, indigenous people usually mean those human groups who live in the same homeland for some generations, have distinct language, custom, and other cultural behavior, usually live a different way of life, and have a lower technology level than the dominant groups in the same country or territory.[3] The estimated five to six thousand indigenous groups spread across the globe, with the heaviest concentration in tropical rainforests. However, compared to 500 years ago when Europeans started to colonize the whole world, the number of indigenous groups has decreased significantly. They used to occupy most of the earth's land surface, but now only live on 12 percent of it; their populations are estimated to be 250 million.[4]

Many indigenous groups still rely only or heavily on resources on the land to sustain their lives. In addition, land also provides indigenous peoples with cultural identity in relation to their history, legend, religion, ceremony, and meaning of life. Without life-supporting and culture-embedding land and the trees, crops, flowers, creeks, and mountains, indigenous people will no longer be able to connect to the natural world, hence they will cease to be indigenous people.

Before the industrial revolution, most indigenous peoples were able to maintain a relatively harmonious relationship with their surrounding environment, mainly because of their low population density, minimum technology use, and simple lifestyle.[5] However, with the global advance of industrialism and capitalism, their way of life and their very survival are constantly under serious threat. In recent years, international communities have gradually come to realize the importance of protecting indigenous people's way of life both because of human rights concern and the significant contribution of indigenous people to environmental protection and ecological knowledge. Notwithstanding these concerns, indigenous communities are still under heavy assault from expansionist capitalist systems.[6]

The demise of the world's indigenous people means not only the destruction of precious cultural heritage, but also the possibility of losing important ecological knowledge forever. Currently, many indigenous groups still claim their homeland in the world's remaining rainforests, including South and Central America, tropical Africa, South Asia, Southeast Asia, and the Pacific islands. While these tropical rainforests encompass only 6 percent of the earth's land surface, it is estimated that more than 50 percent of the world's species are in these areas; hence, they are the world's most important biodiversity regions.[7] Over the past few decades, the link between indigenous people and tropical rainforests has resulted in indigenous groups becoming "guardians of the land."[8] Time and again, their presence and resistance movement has prevented destruction of the environment by outsiders. Similarly, in many "remote" parts of the world, indigenous people have preserved local ecology with their culture and way of life.

It has been well documented and researched that because of their long interaction with the environment surrounding their communities, indigenous people often hold rich knowledge about the local environment. This knowledge, including weather, geographic formation, living organisms, and medicinal plants, is often preserved in their language, custom, and daily life. Because of this, *Our Common Future* affirms the importance of indigenous people in providing beneficial lessons in the "management of resources in complex forest, mountain, and dry land ecosystems."[9] Since most indigenous people do not have a written language, their ecological knowledge is not easily known or passed on to outsiders. Therefore, with the gradual disappearance of the indigenous people, much precious ecological knowledge will be lost forever. Furthermore, the loss of indigenous people often brings about economic activities into the area, hence the destruction of local ecology.

While the demise of the world's diverse species has attracted mounting international concern, indigenous people's similar fate seemed to arouse less consideration. This is mainly due to the lack of understanding about the aforementioned relationship between indigenous people and the environment. As we can still witness today, whereas most "modern societies" have greatly altered or destroyed their immediate environment, most indigenous groups have maintained a relatively healthy ecological state of their homeland. Crosby noted that the expansion of European societies since five hundred years ago was made possible only by converting the world's diverse environment into a similar, Europe-like environment, hence the massive reduction of the world's biodiversity.[10]

To be sure, with many human groups committing to settled agriculture more than 1,000 years ago, the reduction of biodiversity on the earth's surface also began. With the coming of industrialization, capitalist globalization, and "free trade" doctrine, world division of labor has steadily intensified, which further encouraged the single-purpose land use pattern. Examples include massive monocropping in the United States and Canada, and Taiwan's converting farmland and coastal wetlands to industrial usage under export-oriented economic growth policy. All of this can only result in rapid reduction of the world's biodiversity.

It is important to recognize that over the past century, the problem of biodiversity loss has been caused mainly by resource extraction, industrialization (and its accompanying activities), urbanization, agriculture intensification, and expansion of ranching activities. In other words, it is a production- and culture-specific problem; not a problem resulted from all human activities. As mentioned earlier, the highest concentration of biodiversity on the earth's land surface is often the homeland of indigenous peoples. This is mainly because over their long settlement history, most indigenous peoples have not greatly altered their surrounding environment; they depended, managed, and in general, "co-evolved" with their surrounding natural environment.[11] While we should not romanticize indigenous peoples and over-exaggerates their "ecological wisdom,"[12] it is a serious mistake to not distinguish them and their activities from the rest of the world, particularly those engaging in intensified agriculture and industrial activities.

Research evidence has increasingly shown that there is a close relationship between biodiversity and cultural diversity.[13] In its detailed study of 900 ecological regions, World Wildlife Fund International identified 238 most important biodiversity regions; these regions affirm this relationship.[14] Because of these findings, we have come to understand that human groups play different roles in nature's ecosystem; this is especially true for the world's indigenous peoples. In Taiwan, the indigenous peoples have a long history of coexistence with their surrounding environments; today they often work with environmental NGOs to preserve their environment while getting some economic benefits.

Indigenous tourism and ecotourism

Since the early 1990s, indigenous peoples of the world have begun to engage in tourism development that aims not only to help them economically but also to protect the natural environment of their homeland. In the past, indigenous tourism focused on presentation of indigenous groups' distinct cultural heritage.[15] An outstanding example is tourism development among the Pueblos of New Mexico, USA.[16] Since the early 1990s, with the role of indigenous peoples in environmental conservation recognized at the Earth Summit, many indigenous groups gained a new comparative advantage in tourism development—the natural environment that many indigenous homelands are surrounded by.

It may be partly true that tourists from the mainstream society interested in visiting indigenous communities demonstrate their "imperialist nostalgia."[17] Nash goes even further in calling tourism a form of imperialism, especially those tourists coming from wealthy countries to visit less well-off countries and areas.[18] Still, indigenous communities can benefit economically and culturally if they are able to gain control and autonomy in tourist activities.[19] Lujan found that Taos Pueblo was able to maintain the integrity of its culture even with tens of thousands of visitors visiting their community yearly; the reason for this is mainly because of the community's historical heritage, religious strength, and a set of regulations.[20]

Smith identified four Hs—Habitat, Heritage, History, and Handicrafts—as the main elements which indigenous communities can attract tourists of the modern day, and by careful design, these communities "could build upon their assets, to create the type of tourism compatible with local interests and abilities or choose to forgo tourism and seek alternative economic opportunities."[21] Indeed, these four elements have become precious elements in a globalized world where built environment replaces natural environment, and previously distinct cultures are increasingly becoming similar. Indigenous communities, on the other hand, are often located in more pristine natural habitats, and many still maintain their distinct cultural tradition often exemplified in their handicrafts.

Some indigenous communities in Taiwan have long been prime destinations of cultural tourism. However, in the past, most of these communities suffer from lack of autonomy and control over tourist activities.[22] It is only since the late 1990s that increasing numbers of indigenous communities have made better use of their four Hs in promoting community-controlled tourism development. A case in point is the Shanmei (山美) community and its Danayigu (達娜伊谷) Ecological Park. In 1989, when most Taiwanese people knew very little about "ecological conservation" and "sustainable development" concepts, the Tsou people (鄒族) of Shanmei village in the Ali Mountain agreed upon a Danayigu Ecological Conservation Project, and collectively drafted a self-restraint agreement on fishing in their village meeting.[23] Before this meeting, the ecology of Danayigu Creek had been degraded, and fish stock reduced due to ecological degradation, as well as people using electric devices and poison to catch fish. The conservation project was an initiative made by community members to protect and restore the degraded creek in their territory. Village head Gau Cheng-Sheng (高正勝) zealously persuaded many clan elders to release their personal/clan lands to be included in the conservation area and a community task force was subsequently set up to patrol the creek for the protection of fish. After four years of cooperative community work, the fish stock in the creek slowly returned, and the community officially established the "Danayigu Ecological Park" in 1995. The park was opened to tourists with an entry fee for fish viewing, and the community received NT$800,000 for the first year, and the amount more than doubled to reach NT$1,660,000 the next year.

In Gau Cheng-Sheng's discourse, Danayigu is a sacred place of the Tsou people; it has to be well protected by the Tsou people:

> Danayigu is an ecological protection area handed down by our ancestors; we can neither hunt nor fish in this area, no killing whatsoever in Danayigu is allowed anyway. According to our legend, Danayigu is home to the ecological guarding god.[24]

With Shanmei village becoming rich and famous, many indigenous communities hope to follow the same path. Of course, the "Shanmei model" is not without its own problem. H. C. Tang on the one hand recognizes Shanmei's achievement, on the other hand he points out some problems faced by Shanmei: miscommunication between community and government, insufficient professional personnel, and insufficient popular participation of public affairs.[25] Moreover, in recent years, with the increasing number of tourists, tourism in Shanmei has grown from small scale to mass scale. The roads leading to Shanmei are often jammed with tour buses, and tin houses and stands along the roadside have mushroomed; these have greatly disrupted this previously beautiful and quiet small mountain village.

As such, it is important for a community to constantly review its tourism development to ensure better community control and adaptability. Recent literature on ecotourism offers good guidelines for indigenous communities to reflect on. Specifically, Honey identifies seven characteristics that communities should demonstrate in order to have their operation being considered as ecotourism: (1) Involves travel to natural destinations; (2) Minimizes impact; (3) Builds environmental awareness; (4) Provides direct financial benefits for conservation; (5) Provides financial benefits and empowerment for local people; (6) Respects local culture; and (7) Supports human rights and democratic movements.[26] These characteristics involve one dimension and two actors; the dimension of appreciation and preservation of ecological beauty, and tourists and local communities as actors. In other words, local community and residents (the host) are connected to outside tourists (the guest) through local habitat; ecotourism emphasizes the need to make such connection beneficial to local community, its natural environment, and visiting guests in terms of educational purpose.

Indigenous communities and NGOs

It has been well documented that the progress of ecological conservation in Taiwan since the 1980s can mostly be credited to the works of environmental NGOs. Similarly, over the past 20 years, many indigenous communities have gained popular recognition with their ingenious community-based conservation initiatives. However, not all indigenous communities have the capacity, know-how, human power, and capital to institute their conservation projects. As such, some environmental NGOs have come to assist indigenous communities in a joint effort to conserve local environment, while promoting job security and cultural continuity. In Hualien County, with high concentration of indigenous population, a few environmental groups have worked closely with indigenous communities to promote nature conservation-related local initiatives.

The Wilderness Protection Association Hualien chapter (WPAH), for example, was very active in opposing the government's highway widening project and a proposed power plant in the late 1990s, but in recent years, it has helped different levels of local governments in natural resource inventory research, in ecological interpretation training, and in designing eco-cultural tourism of an indigenous fishing village. Similarly, the Environmental Protection Alliance Hualien chapter (EPAH) was involved in the government's tourism enhancement project in the indigenous communities of Fuli (富里) and Shuilien (水璉). In its Shuilien project, for example, EPAH helped the community to obtain government funding for reestablishing habitat for wild lilies, and for redesigning and landscaping the community in order to attract tourists. It is worth mentioning that both groups tried to work closely with local communities with an aim to empower these communities so that communities themselves would be better prepared to engage in tourism development.

The WPAH chairperson indicates:

> We have to acknowledge that many communities are interested in tourism development. We help them with training in environmental interpretation so that their communities could develop ecological tourism, and ultimately,

> we hope that through our helping these communities, we are able to better protect our environment.

Local NGOs' close work with communities in recent years, was in many ways facilitated by government's "community revitalization project" (operating since the early 1990s). Both the Council for Cultural Affairs of the central government and the Cultural Bureau of local government are eager to help communities in developing ways to rebuild their social fabrics. NGOs quickly became the government's partners or mediating organizations, with ecologically sound tourism development projects being the favored goal.

In Dalanfu (達蘭埠) community, a small indigenous Amis (阿美族) community located in southern Hualien County with about 200 residents, was recently made famous with its production of day lily, a delicacy to many Taiwanese people. In early 2000s, with the help of World Vision Hualien Chapter (WVHC), many community members, centered on church leaders, decided to switch to organic farming for the health of both the local environment and the human body. It was a very successful move economically. While day-lily flowers as food must be picked before their full blossom; a sea of day-lily flowers makes a perfect picture. Local people thus decided to venture into ecotourism by leading tourists into their day-lily field, and further to their remote traditional settlement.

WVHC has been working with many indigenous communities in Hualien area for a few decades, and has gained trust as well as confidence among indigenous people and communities. As most Dalanfu community members go to the same Christian church it was easy for them to form consensus centering on the church. As such, when WVHC came up with the idea of organic farming of day lilies and gained support of key church members including the minister, other members followed suit quickly and, in less than two years, most community members were in the business of organic day-lily farming, which also helped promote the image of the community's product as a healthy and environmentally friendly product.

Located in the coastal area of southern Hualien, Gangkou (港口) community is inhabited by seafaring Amis people. Traditionally, they build light rafts using locally available and sustainable materials—bamboo. However, since the 1980s, with the readily available and durable plastic pipes, bamboo rafts were gradually being replaced by plastic rafts, and the younger generation has lost the skill to build bamboo rafts. In trying to encourage Gangkou villagers' use of sustainable materials, and to pass on the skills of bamboo raft making, the Kuroshio Ocean Educational Foundation—a Hualien-based organization with major interest in ocean education—applied for a government grant to work with community members in building bamboo rafts. The project was very successful in recruiting both the older generation that still have the knowledge of raft building, and the younger generation that came to learn the skills of bamboo raft building from their elders.

Discussion and conclusion

Over the past ten years, many NGOs have worked with indigenous communities to promote local sustainable development and ecotourism in Hualien County. In particular, the Wildness Society Hualien Chapter has helped many indigenous communities in their ecotourism interpreter training. The Kuroshio Ocean Educational Foundation worked with the Gangkou community to build traditional rafts from local, sustainable materials (in place of modern-day plastic pipe rafts). The World Vision Hualien Chapter worked with the Dalanfu community in their organic farming and ecotourism development. The Taiwan Environmental Protection Union Hualien Chapter worked with many indigenous communities in pollution prevention, indigenous culture promotion, and in ecotourism and community forestry development. A list of NGOs working with indigenous communities in Hualien is shown in Table 6.1.

In general, this research finds that indigenous communities in Hualien County welcome NGOs working with them. Dalanfu has benefited greatly economically with World Vision's help in promoting organic day-lily farming, while Fuli community welcomed the same organization's support in mapping their traditional homeland and in ecotourism interpreter training. In Shuilien community on the east coast, Taiwan Environmental Protection Union Hualien Chapter joined local people in opposing the construction of a coal-burning power plant; if built, the plant would seriously pollute the land, air, and ocean. In addition, many community members have praised NGO members

Table 6.1 NGOs working with indigenous communities in Hualien County

Name of NGO	*Major works*
Taiwan Environmental Protection Union Hualien Chapter	Pollution prevention, ecological conservation
Kuroshio Ocean Educational Foundation	Marine ecology protection, marine culture promotion
Wild at Heart Association Hualien Station	Nature conservation, Indigenous community's environmental protection
Wildness Society Hualien Chapter	Nature conservation
Hualien Sunrise Environmental Group	Pollution prevention, ecological conservation
Mennonite Social Welfare Foundation	Social service, indigenous health, and social welfare
World Vision Hualien Chapter	Indigenous community development
Legal Aid Foundation Hualien Chapter	Indigenous legal consultation, community support
Various Community Development Association	Community development

Source: Compiled by the author.

for their cultural sensitivity and their appreciation of indigenous culture and custom.

Notwithstanding the aforementioned positive aspects of NGOs and indigenous communities working together, many obstacles still exist in indigenous communities' move towards sustainable development. Some NGO members indicate that many indigenous communities are quite divided politically, preventing community members from working well together. Partly because of this, some communities depended heavily on outside help, both governmental and non-governmental, in promoting community development-related affairs. Outside help could be beneficial for the short term, but in the long run, community members have to work together by themselves to build up their economic base, preserve their community environment, and to promote cultural preservation and development. In Full community for example, while some community members have worked hard to build viable small-scale ecotourism operations, the operation has been restricted in recent years due to in-fighting among different clans and some individuals.

One NGO worker also indicated that many indigenous communities have put much emphasis and resources on tourism development, to the extent that other community socio-cultural-environmental agendas have been downplayed or neglected. While ecotourism development in indigenous communities of Hualien is certainly one of the major ways to approach local sustainability, there are other equally important community issues to be taken care of. In addition, ecotourism development requires thorough planning and training of personnel; without good preparations, ecotourism may not bring either economic or socio-environmental benefits to the community.

In sum, this research found that while indigenous communities-NGOs cooperation in Hualien has borne much sweet fruit, the communities and the operation are confronted with the following challenges:

1 Community internal disputes constitute major obstacles for community members to support coordinated community projects and activities.
2 Due to legal and government administrative limitations, most communities cannot have effective control over visiting guests; this has resulted in the difficulties for communities to achieve autonomy and to have better management of and planning for ecotourism operation.
3 Most communities have yet to come up with viable benefit sharing mechanisms; this is one of the major factors contributing to local disputes. Having adequate benefit sharing mechanisms is another key to overall success in indigenous communities' sustainable ecotourism development.
4 Finally, many communities studied in this project are still in their initial stage of ecotourism and organic farming project development; there is still much room for improvement in promoting indigenous traditional ecological knowledge so that they are incorporated into these operations.

Notes

1 H. F. French, "Forging a New Global Partnership," in *State of the World 1995*, (New York: W. W. Norton & Company, 1995), p. 170.
2 J. A. McNeely et al., *Conserving the World's Biological Diversity*, (Gland, Switzerland: IUCN, 1990), p. 17.
3 A. T. Durning, *Guardians of the Land: Indigenous Peoples and the Health of the Earth*, Worldwatch Paper 112, (Washington, D.C.: Worldwatch Institute, 1992), pp. 1–62.
4 S. Davis, "Introduction," in Shelton Davis (ed.), *Indigenous Views of Land and the Environment*, World Bank Discussion Papers, No. 188, (Washington, D.C.: World Bank, 1993), p. 1; Durning, *Guardians of the Land*, pp. 1–62.
5 K. Milton, *Environmentalism and Cultural Theory: Exploring the Role of Anthropology in Environmental Discourse*, (London: Routledge, 1996), pp. 106–141.
6 C. C. Chi, "From Guest to Host: Toward Indigenous People's Control of Tourist Activities in Taiwan," in *Proceedings of Indigenous Culture and Tourism Development Conference*, (Taipei: The Outdoor Recreation Association of R.O.C. 1998), pp. 21–36.
7 E. O. Wilson, *The Diversity of Life*, (London: Penguin books, 1992), p. 185; McNeely et al., *Conserving the World's Biological Diversity*, pp. 83–106.
8 Durning, *Guardians of the Land*, pp. 1–62.
9 The World Commission on Environment and Development, *Our Common Future*, (Oxford: Oxford University Press, 1987), p. 12.
10 A. W. Crosby, *Ecological Imperialism: The Biological Expansion of Europe, 900–1900*, (Cambridge: Cambridge University Press, 1986), pp. 294–311.
11 L. Maffi et al., *Indigenous and Traditional Peoples of the World and Ecoregion Conservation: An Integrated Approach to Conserving the World's Biological and Cultural Diversity*, A WWF Report, (Gland, Switzerland: WWF International, 2000), p. 13; E. A. Smith, "On the Coevolution of Cultural, Linguistic, and Biological Diversity," in Luisa Maffi (ed.), *On Biocultural Diversity*, (Washington, D.C.: Smithsonian Institution Press, 2001), pp. 95–117.
12 Milton, *Environmentalism and Cultural Theory*, pp. 109–114.
13 Durning, *Guardians of the Land*, pp. 1–62; L. Maffi, "Introduction: On the Interdependence of Biological and Cultural Diversity," in Luisa Maffi (ed.), *On Biocultural Diversity*, (Washington, D.C.: Smithsonian Institution Press, 2001), pp. 1–50.
14 Maffi et al., *Indigenous and Traditional Peoples of the World and Ecoregion Conservation*, pp. 1–88.
15 V. L. Smith, "Introduction," in Valene L. Smith, (ed.), *Hosts and Guests—The Anthropology of Tourism*, (Philadelphia: University of Pennsylvania Press, 1989), pp. 1–20.; R. E. Wood, "Touristic Ethnicity: A Brief Itinerary," *Ethnic and Racial Studies* 21.2 (1998): 218–241.
16 C. C. Lujan, "A Sociological View of Tourism in an American Indian Community: Maintaining Cultural Integrity at Taos Pueblo," *American Indian Cultural and Research Journal* 17(3): 101–120; M. J. Martinetz, *Double Take: Tourism and Photography Endeavors among the Northern Pueblos of the Rio Grande*, (Ph.D. Thesis, University of Minnesota, USA, 2008), pp. 1–214.
17 R. Rosaldo, "Imperialist Nostalgia," *Representations* 26 (1989): 107–122.
18 D. Nash, "Tourism as a Form of Imperialism," in Smith (ed.), *Hosts and Guests*, pp. 37–52.
19 Lujan, "A Sociological View of Tourism in an American Indian Community: Maintaining Cultural Integrity at Taos Pueblo," pp. 101–120; Chi, "From Guest to Host," pp. 21–36.
20 Lujan, "A Sociological View of Tourism in an American Indian Community: Maintaining Cultural Integrity at Taos Pueblo," pp. 101–120.

21 V. L. Smith, "Introduction," p. 288.
22 H. C. Hsieh, *'Mountain People Tourism': An Anthropological Interpretation of the Presentation of Contemporary Mountain Culture*, (Taipei: Chi-Li Evening Post, 1994), pp. 145–166.
23 Y. C. Wen, "Danayigu Ecological Conservation and Community Building," *Shan Hai* 18 (1998): 59–63.
24 C. S. Kau, *Walking through Gloomy Valley: The Legend of Danayigu*, (Taipei: Taiwan Presbyterian Church, 2005), p. 44.
25 H. C. Tang, *Sustainable Community Development of the Tzou Nation: The Case of Sanmei Community in Ali Municipality*, (Master Thesis, Graduate Institute of Social Welfare, National Chong Chen University, 2002), pp. 1–202.
26 M. Honey, *Ecotourism and Sustainable Development: Who Owns Paradise?* (Washington, D.C.: Island Press, 1999), pp. 22–24.

Bibliography

Chi, C. C. "From Guest to Host: Toward Indigenous People's Control of Tourist Activities in Taiwan," in *Proceedings of Indigenous Culture and Tourism Development Conference*. Taipei: The Outdoor Recreation Association of R.O.C. 1998, pp. 21–36.

Crosby, A. W. *Ecological Imperialism: The Biological Expansion of Europe, 900–1900.* Cambridge: Cambridge University Press, 1986.

Davis, S. "Introduction," in Shelton Davis (ed.), *Indigenous Views of Land and the Environment*, World Bank Discussion Papers, No. 188, Washington, D.C.: World Bank, 1993.

Durning, A. T. *Guardians of the Land: Indigenous Peoples and the Health of the Earth*, Worldwatch Paper 112. Washington, D.C.: Worldwatch Institute, 1992.

French, H. F. "Forging a New Global Partnership," in *State of the World 1995*. New York: W. W. Norton & Company, 1995.

Honey, M. *Ecotourism and Sustainable Development: Who Owns Paradise?* Washington, D.C.: Island Press, 1999.

Hsieh, H.C. *'Mountain People Tourism': An Anthropological Interpretation of the Presentation of Contemporary Mountain Culture*, Taipei: Chi-Li Evening Post, 1994.

Kau, C. S. *Walking through Gloomy Valley: The Legend of Danayigu*. Taipei: Taiwan Presbyterian Church, 2005.

Keller, R. H. and M. F. Turek. *American Indians and National Parks*. Tucson: University of Arizona Press, 1998.

Lujan, C. C. "A Sociological View of Tourism in an American Indian Community: Maintaining Cultural Integrity at Taos Pueblo," *American Indian Cultural and Research Journal* 17 (1993): 101–120.

Maffi, L. "Introduction: On the Interdependence of Biological and Cultural Diversity," in Luisa Maffi (ed.), *On Biocultural Diversity*. Washington, D.C.: Smithsonian Institution Press, 2001, pp. 1–50.

Maffi, L., G. Oviedo, and P. B. Larsen. *Indigenous and Traditional Peoples of the World and Ecoregion Conservation: An Integrated Approach to Conserving the World's Biological and Cultural Diversity*. A WWF Report. Gland, Switzerland: WWF International, 2000.

Martinez, M. J. *Double Take: Tourism and Photography Endeavors among the Northern Pueblos of the Rio Grande*. Ph.D. Thesis, University of Minnesota, USA, 2008.

McNeely, J. A., K. R. Miller, W. V. Reid, R. A. Mittermeier, and T. B. Werner. *Conserving the World's Biological Diversity*. Gland, Switzerland: IUCN, 1990.

Milton, K. *Environmentalism and Cultural Theory: Exploring the Role of Anthropology in Environmental Discourse*. London: Routledge, 1996.

Minister of Indian Affairs and Northern Development. *The Champagne and Aishihik First Nations Final Agreement Implementation Plan*. Ottawa, 1993.

Nash, D. “Tourism as a Form of Imperialism,” in V. L. Smith (ed.), *Hosts and Guests: The Anthropology of Tourism*, 2nd edition, Philadelphia: University of Pennsylvania Press, 1989, pp. 37–52.

Rosaldo, R. “Imperialist Nostalgia,” *Representations* 26 (1989): 107–122.

Shiva, V. *Biopiracy*. Boston: South End, 1997.

Smith, E. A. “On the Coevolution of Cultural, Linguistic, and Biological Diversity,” in Luisa Maffi (ed.), *On Biocultural Diversity*. Washington, D.C.: Smithsonian Institution Press, 2001, pp. 95–117.

Smith, V. L. “Introduction,” in Valene L. Smith (ed.), *Hosts and Guests—The Anthropology of Tourism*. University of Pennsylvania Press, 1989, pp. 1–20.

Smith, V. L. “Indigenous Tourism: The Four Hs,” in Richard Butler and Thomas Hinch London (eds.), *Tourism and Indigenous Peoples*. London: International Thomson Business Press, 1996, pp. 283–307.

Spence, M. D. *Dispossessing the Wilderness: Indian Removal and the Making of the National Parks*. Oxford: Oxford University Press, 1999.

Tang, H. C. *Sustainable Community Development of the Tzou Nation: The Case of Sanmei Community in Ali Municipality*. Master Thesis, Graduate Institute of Social Welfare, National Chong Chen University, 2002.

Taroko National Park. *Bringing People Back In: Local Participation and Natural Resource Management*. Hualien: Taroko National Park, 2001.

Wen, Y. C. “Danayigu Ecological Conservation and Community Building,” *Shan Hai* 18 (1998): 59–63.

Wilson, E. O. *The Diversity of Life*. London: Penguin books, 1992.

Wood, R. E. “Touristic Ethnicity: A Brief Itinerary,” *Ethnic and Racial Studies* 21.2 (1998): 218–241.

World Commission on Environment and Development. *Our Common Future*. Oxford: Oxford University Press, 1987.

Part III

Case studies of Japan

7 Pyroclastic rivers

The Hôei Fuji Eruption (1707)

Beatrice M. Bodart-Bailey

Introduction

In the morning of 23.11.Hôei (寶永) 4 (16.12.1707), accompanied by loud sounds of explosions, there arose from Mount Fuji (富士山) a column of what looked like belching black smoke rising to an altitude of some twenty kilometers.[1] It was, however, not just smoke but mostly volcanic fall-out or tephra, containing rocks and pebbles of various sizes, and marked the onset of the volcano's most violent eruption in recorded history. The event has also been described as "the worst ash fall disaster in Japanese history,"[2] yet it finds scant mention in Japanese history books, perhaps because it was followed neither by riots nor famine like later eruptions.[3] Further, one would have to admit that the government of the fifth shogun Tokugawa Tsunayoshi (德川綱吉, 1646–1709) dealt efficiently with the immediate consequences, an action at variance with the traditional criticism of his administration.[4] But while historians have much ignored the event, volcanologists and environmentalists have shown great interest, especially recently. The pattern of previous eruptions suggests that now, over three hundred years after the last, a further eruption is overdue. Attempting to learn from the 1707 eruption, historical sources describing the process have been examined and hazard maps predicting the scale of future eruptions have been drawn.[5] The Great East Japan Earthquake of March 2011 further drew attention to the Hôei eruption prompting articles to appear even in popular journals, for that eruption too was preceded by major seismic activity.[6] Finally Fuji's nomination as World Heritage Site in April 2013 generated new attention to Japan's iconic mountain and its past activities.

To begin with, this chapter shows the dimensions of the eruption using contemporary reports of the event together with the scientific analysis of volcanologists. It then examines the response of bakufu (幕府) to the situation, relying chiefly on records left by the locally affected population, containing details not found in the official government record. Unlike his predecessors and successors, the fifth shogun had not only proclaimed the need for "benevolent rule (*jinsei* 仁政)," but repeatedly punished officials who went against that principle. Local documentation shows that the farmers were well aware of this fact, and skillfully used it to negotiate assistance from their domain lord. As the bakufu became aware of the scale of the disaster, an unprecedented tax was levied on the rest of the country to provide relief, and the areas affected were placed directly under its control. However, with the death of the fifth shogun, other financial priorities came to prevail under his successor and the two officials originally in charge died under unexplained circumstances.

The eruption

Fuji's earliest documented eruption took place in 781, with major eruptions in 800 and 864. Several events followed in the tenth and eleventh centuries, and further eruptions were recorded in 1511 and 1560, but they had all been minor.[7] However, as the seventeenth century was drawing to a close, Japan was plagued by an increasing amount of seismic activity resulting in a number of strong earthquakes. On 2.11.Genroku (元禄) 16 (10.12.1703) an earthquake of 8.2 magnitude and accompanying tsunami caused enormous devastation and loss of life in Edo (江戸) and the surrounding coastal area. The epicenter was close by, in Sagami Bay (相模灣), but Fuji remained quiet.[8] Four years later, on 4.10.Hôei 4 (28.10.1707) an earthquake of magnitude 8.7 with its epicenter to the south of the Kii (紀伊) peninsula caused major damage in Nagoya, and also shook the villages around Mount Fuji. From then on tremors could be felt daily in the Fuji area, increasing in strength until the violent explosive activity forty-nine days later.[9]

This time the eruption occurred not from the central vent of the summit crater as previously, but was a flank eruption half-way down the southeastern side of the volcanic edifice. Three new craters were formed plus a hump, destroying forever the perfect cone-shaped structure that had resulted from the symmetrical accumulation of pyroclastic material of previous eruptions through the central vent. When the eruption came to an end after sixteen days and the clouds of smoke and debris cleared, the hump became visible and was named Mount Hôei after the era name[10] (see Figure 7.1).

As shown in Figure 7.1, the inscription on this painting does not provide details of the time, but only states that in the morning of 23.11.Hôei 4, upon an earthquake Fuji roared like thunder and fire emerged as shown. It notes: "From the cloud of fire, stones and sand fell, like big stars. They piled up to form Mount Hôei. However, on 23.11. this could barely be seen."[11]

The craters that formed at the time are known as Hôei I, II, and III, situated roughly between 3,100 meters and 2,100 meters above sea level with Hôei I being by far the largest and situated at the highest altitude.[12]

Eruption damage

Though the soil on the lower slopes of Mount Fuji yielded only relatively sparse agricultural product, the area was settled by a number of villages. An important route linking the coastal area with the inland provinces traversed the area and the villagers supported themselves by providing accommodation and pack horses for

Figure 7.1 The painting depicts the eruption as seen from the vicinity of Numazu in Shizuoka (source: National Museum of Japanese History (ed.), *Documenting Disaster: Natural Disasters in Japanese History, 1703–2003* (2003), p. 60).

travelers. Moreover, since ancient times Mount Fuji had been the object of religious veneration and some locals made a living as guides (*oshi* 御師) for those wishing to climb the steep slopes to the summit. One of these guides, Tsuchiya Idayû (土屋伊太夫), left a record of the unprecedented eruption. A number of other records exist, but as mountain guide, Idayu's knowledge of the area and attention to environmental conditions was acute, and his description is hence considered to be reliable and especially valuable.[13]

Strong earthquakes had been felt the previous day and also in the morning of 23.11.Hôei 4. Idayû recorded that at around 10 a.m. the mountain emitted such loud noises that residents feared it would break up. At the same time the sky turned black, and then rocks the size of tea kettles or large tenmoku (天目) bowls rained down from heaven. Breaking on impact into four to five pieces, they emitted flames, setting on fire the thatched roofs of houses. There was a lull around three o'clock in the afternoon, but in the evening it started to rain rocks again and by morning they had accumulated to a depth of some seventy-five centimeters (two *shaku* 尺 five *sun*寸) Idayû noted.[14] However, this was only the beginning of some sixteen days of volcanic activity, and by the end, deposits of three to four meters had accumulated in the south-western area at the foot of Mount Fuji.

Surveys of the deposit have indicated that during the first hours of the eruption pumice and obsidian were ejected from the second and third craters causing a preliminary fall-out of whitish material, followed by material of basaltic composition.[15] The lighter particles became airborne and were blown by the prevailing seasonal winds east-north-east, reaching the city of Edo (today's Tokyo) located at a distance of some 100 kilometers.

In the morning of the first day of the eruption, 23.11.Hôei 4, the daimyo (大名, great territorial lords) were assembling at Edo castle for the usual confirmation and awarding of titles. Attending the ceremonies, the monk Ryûkô (隆晃, 1649–1724) noted in his diary that sliding doors and screens (*shoji* 障子) at his residence and in the castle rattled as if shaken by wind or an earthquake, but there was neither. Other diarists similarly noted this strange occurrence.[16] Today scholars attribute this unusual phenomenon to the shock waves of the explosion reaching Edo with the prevailing winds.[17] Eventually, tremors began to be felt, black clouds appeared in the sky and it became dark as if night was falling. By three o'clock in the afternoon it was raining white ashes, covering the city like snow.[18]

The nine-year-old son of Yabuta Shigemori (藪田重守, 1664–1747), a senior retainer of the Grand Chamberlain Yanagisawa Yoshiyasu (柳澤吉保, 1658–1714), was intrigued by this unprecedented phenomenon, and carefully preserved some of the material that rained from the sky that day. Two small packets of roughly ten grams of material each came to the attention of scholars in 1996 and the contents has been carefully analyzed. An accompanying note states that the ash was collected after midday on 23.11.Hôei 4 as thunder could be heard. The location is believed to have been at Uchikanda (内神田) 1 chôme, to the northeast of Kandabashi crossing (神田橋交差点), some 94 kilometers from the source of the ashes: Hôei craters II and III.[19] With the larger part of the sample examined consisting of grains between 0.25 and 0.063 millimeter in diameter, the analysis confirmed previous research indicating that upper level winds were blowing from the WNW carrying the finest particles toward Edo. At a lower level, winds were blowing from the WSW depositing larger-sized material in areas further south as shown in Figure 7.2. Thus records from Fujisawa (藤沢) and Enoshima (江の島) speak of it raining stones, rather than the ash or sand deposited further north. The analysis also confirmed the established sequence of the eruption, of dense deictic white to grey ash being deposited the first day with darker scoria falling later.[20]

In Edo people wondered whether Mount Asama (浅間山) to the northeast of the city had erupted; however, by the evening of the next day, 24.11.Hôei 4, a messenger arrived in Edo from the Tôkaidô (東海道) station of Yoshiwara (吉原) some twenty-five kilometers south of Mount Fuji, and reported the location of the eruption and the great damage it was causing, burying villages and fields under ashes and sand.[21] The next morning the shogun ordered a group of three inspectors (*okachi metsuke* 御徒目付) to depart for a survey of the stricken area.[22]

The shogunal inspectors arrived at Subashiri (須走, present-day Oyamachô 小山町), the village closest to the location of the eruption, on 28.12.Hôei 4 as the fire at the source of the eruption was still burning at a distance of some ten kilometers. The area was filled with smoke and visibility was bad. Only the tall

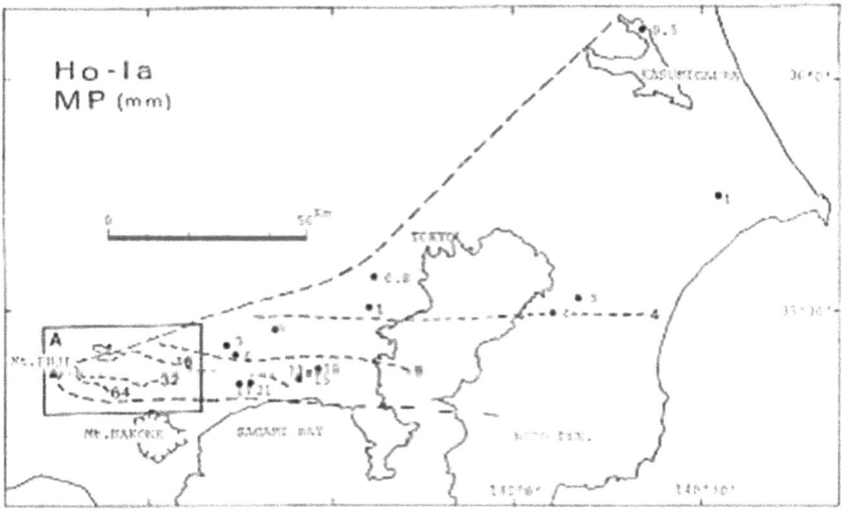

Figure 7.2 Maximum diameter in mm of grain deposited on 23.11.Hôei 4 (source: Miyaji Naomichi 宮地直道 and Koyama Masato 小山真人, "Recent Studies of the 1707 (Hôei) Eruption of Fuji Volcano (富士火山1707年噴火（宝永噴火）についての最近の研究成果)," *Fuji Kazan* 富士火山, 2007, p. 341).

gables of the village's shrine and of some of the larger houses were still protruding from the fall-out. The remaining houses had been totally buried under ashes and rocks of various sizes and the village was deserted. Even at that early date the inspectors estimated the deposit at some three meters. The trees in the area were stripped of their leaves and burnt. Wells and rivers were similarly buried under the rubble, and so were, of course, any fields.[23]

Continuing for sixteen days, the eruption caused considerable damage to the environment. On the basis of contemporary reports, scholars have mapped the distribution of fall out that formed an elliptical shape oriented in an easterly direction. As Figure 7.3 shows, even areas some fifty kilometers from the source were covered with deposits of tephra up to thirty centimeters and a large part of present-day Kanagawa (神奈川) prefecture, including Tamachi (田町), Fujisawa, Enoshima and even Yokohama (横浜), had a covering of at least sixteen centimeters deep. Examining the impact of tephra on plants and soils, the New Zealand Ministry of Agriculture and Forestry defines a 2.5–15 centimeter layer of ash as "deep burial" and concludes that "Vegetation canopy recovery takes decades." Moreover, even light coverings of tephra may distribute harmful chemicals like fluorine, polluting drinking water and leading to death from fluorosis, especially with animals as they attempt to graze on any remaining vegetation[24] (see Figure 7.3).

The deposits that had accumulated on the fields at the end of the sixteen-day eruption were only one part of the problem: with the onset of rains the heavy

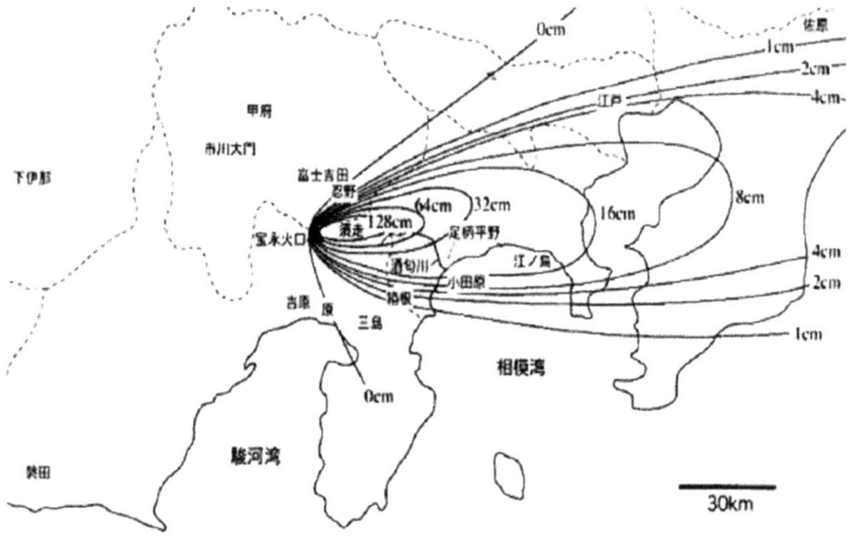

Figure 7.3 Distribution of sediment after sixteen days of eruption (source: Miyaji Naomichi 宮地直道 and Koyama Masato小山真人, "Recent Studies of the 1707 (Hôei) Eruption of Fuji Volcano (富士火山1707年噴火（宝永噴火）についての最近の研究成果)," *Fuji Kazan* 富士火山, 2007, p. 341).

deposits on the mountain slopes were washed into streams and waterways, silting up riverbeds and forming blockages and dams that altered the course of rivers. This caused large-scale flooding, and was particularly detrimental in a country where rice cultivation requires a network of clean water channels. For some forty years successive governments would be challenged with the task of repairing the damage caused by the eruption of 1707.

The documentary record

One problem in dealing with the history of human response to the environment in the Tokugawa period is the documentary record. Not only were the government and historians city-based with limited knowledge of rural areas, but they were also samurai, considering the details of the "commoners" in the country of little importance as long as disturbances did not cause a threat to the central administration. The rural population, on the other hand, often lacked the skills and means to keep detailed records. In case of the 1707 eruption of Mount Fuji, however, a lengthy document known as "Record of Petitions Concerning Sand Falling from Mount Fuji (Fujisan sunafuri sogan kiroku 富士山砂降り訴願記録)" forms the exception. The document is attributed to Suzuki Rizaemon (鈴木利左衛門), the headman (*nanushi* 名主) of Yamakita (山北) village in Kawamura (河村) district, Ashigara-gun (足柄郡).[25] Ashigara-gun, part of the Odawara

(小田原) domain of the senior councilor Ôkubo Tadamasu (大久保忠正, 1656–1713), was one of the worst-hit areas. The document describes in some detail how the local population attempted to cope with the environmental damage. To obtain relief they made full use of the knowledge that the fifth shogun, unlike his predecessors and successors, saw the samurai not just as the ruling class, but as "civil servants" in the true sense of the word. Thus when farmers rose in protest because of harsh impositions, the shogun would punish not the farmers, but the officials, and at times even the domain lord, holding them responsible for letting living conditions deteriorate to such an extent that the farmers were unable to attend to their fields.[26] After the death of the shogun, samurai historians greatly criticized his administration which curbed the privileges of the samurai for the benefit of the commoners, and modern historians relying mostly on samurai records, have generally followed this trend.

The process by which the farmers obtained government assistance after the eruption of Mount Fuji well demonstrates the political strength the commoners attained during the government of the fifth shogun. While on one hand supporting the rural population was in tune with the Confucian concept of benevolent government, his successors considered the farmers' strength to make their demands heard a threat to government authority. As a consequence the shogun's name was blackened, suggesting that his government was irrational and dilettante.[27] The government's speedy response preventing the outbreak of famine did not fit the political picture future governments wanted to hand to posterity and official reports were not preserved. The record of events by the village headman is the only one describing the situation in detail and important not only because it demonstrates the political power the rural population attained under the fifth shogun, but also why the eruption of Mount Fuji was not followed by famines or militant riots such as those that occurred after the eruption of Mount Asama later in the century when the government reverted to a policy of putting down rural uprisings rather than investigating and removing the cause.

The early stage: petitions for assistance to the domain lord

The headman's report starts with a description of the damage after the eruption stating that at the base of Mount Fuji the accumulated tephra was up to 4.5 meters, while at the village of Yamakita the fallout was around some 55–63 centimeters. Hence three villages in the Kawamura area decided to petition the domain government for help.

The first petition addressed to the intendants (*daikan* 代官) of the Odawara domain, Ônishi Kakunouemon (大西覚之右衛門) and Aoki Niemon (青木仁右衛門), was dated the twelfth month of Hôei 4 only, but since it mentioned that the fall-out was still accumulating, it must have been written before the eighth day of that month when the fall-out finally stopped. The petition was signed by the headmen of three villages of the Kawamura area, including the headman of Yamakita, who preserved the documents. Inspectors from the domain had visited earlier, but deposits had since increased and they wanted the domain to re-assess the situation. All crops on the fields as well as all other plants, including the lacquer trees from which the villagers normally garnered revenue, had been destroyed. The farmers were attempting to move the tephra, but as the villages were on mountainous terrain, the deposits from higher areas were constantly being washed down from the uplands. Many homes on the higher reaches had been destroyed and people had nowhere to live. There was not enough drinking water. As previously reported to the domain, the summer's harvest of crops had already been destroyed previously by insects and now people had absolutely nothing to eat and were starving.[28]

By the middle of the twelfth month the domain official Yanagida Kyûzaemon (柳田九左衛門) was sent from Edo to transmit orally the daimyo's reply which was duly recorded. The message reassured villagers that their lord was paying the utmost attention to the situation and would make certain that neither people nor domestic animals would die of starvation. He repeatedly expressed his compassion, and instructed the village headmen and their subordinates to ensure that all people were diligently attending to their usual duties. Concrete details of relief or assistance with removing the fall-out, however, found no mention.[29]

The villagers were not satisfied with such vague promises. In the meantime the protest movement had considerably widened: on 26.12.Hôei 4 (18.1.1708), 104 villages sent their representatives to a gathering at Ushijima (牛嶋) village, strategically located on the Yakurazawa ôkan (矢倉沢往還), a major route linking Edo with the Tôkaidô post station of Numazu. Here they composed a new petition to be handed to Kyûzaemon before his departure for Edo the next day.

The new petition asked that help be provided by the domain to remove the rocks and sand that covered the fields. The farmers argued that they had carefully assessed the labor necessary for the task, and that it was impossible to clear the land without assistance. Three years previously they had to cope with the damage caused by the Genroku earthquake, major floods had devastated their fields in the two following years, and this year already two typhoons had caused great damage. Now their land was covered with stones and sand, there were no supplies left from the previous harvest, and they were dying of starvation.[30] The petition was addressed to Yanagida Kyûzaemon and Ôtsu Zensaemon (大津善左衛門), the domain's magistrate (*bugyô* 奉行) stationed at Odawara.[31]

The farmers obviously had little confidence that the petition would be effective, for on the very day it was handed to Yanagida Kyûzaemon departing for Edo, they gathered at another village to agree on a meeting place immediately after the New Year celebrations on the third day of the first month. On that day it was decided that the 104 villages would combine to stage a mass protest, and that the inhabitants would gather on the fourth at Odawara in preparation for a march on Edo. A note was handed to the magistrate Ôtsu Zensaemon explaining why the villagers had decided on this action: The fields were covered with a heavy layer of sand and stones, and there was no way that the farmers, suffering already from a lack of food, would be able to remove them in time for planting in spring. Villages with the task of supplying pack horses for travelers could no longer do so, since there was no food for the animals. They realized that the domain had suffered great losses in the past, and had reduced taxes in times of hardship so farming could continue. Now, however, there was no way the farmers could continue without help and they were going to depart for Edo the next day to petition the bakufu for assistance.[32]

No doubt the farmers were well aware that their protest at Edo would have dire consequences for the domain administration and especially the officials in charge of the relief operation. The ensuing exchanges show how the farmers capitalized on the fear of the officials to obtain their ends.

Thus when the inhabitants of 104 villages of Ashigara-gun threatened to march to Edo to submit their request, the magistrate Ôtsu Zensaemon did his utmost to stop them. Finally the villagers agreed to wait until 6.1.Hôei 5 (28.1.1708). However, arguing that if there were further delays, the fields might not be cleared in time for the spring planting, the farmers assembled again at Odawara on 7.1.Hôei 5 (29.1.1708) to prepare for departure the next day. The local officials presented them once more with the detailed sentiments of the domain lord, but the villagers said they needed food and were going to march to Edo to request that directly from their lord, and at the same time petition the bakufu for assistance to clear the fields.[33]

At that point the Odawara domain office announced that the two officials Itô Bunnai (伊藤文内) and Hironaka Iemon (広仲伊右衛門) had already proceeded to Edo on 2.1.Hôei 5 (24.1.1708) and 4.1.Hôei 5 (26.1.1708) respectively and that these men would soon be returning with an answer from the domain lord. The domain office strongly urged the farmers' representatives to wait at least until 10.1.Hôei 5 (1.2.1708), and ultimately they agreed.

On hearing that their representatives had agreed to wait at Odawara, their followers decided to take matters into their own hands. When domain officials woke up on 8.1.Hôei 5 (30.1.1708), they discovered that four to five thousand farmers had set out during the night and were on their way to Edo. Five officials including the magistrate Ôtsu Zensaemon chased after them and promised them relief rice of five *gô* (合, 1 *gô*=0.18 liter) for every man and two *gô* for every woman if they would wait another two days. The farmers, however, were not mollified, and with great noise and clamor proceeded on their way. They obviously knew how to strike fear in the hearts of the officials, for a noisy protest march of thousands of farmers would undoubtedly result in heavy punishment for the officials' failure to keep peace in the area. Finally the officials added the concession that the village headmen and their deputies would be permitted to proceed to Edo and managed to persuade the farmers to turn back.[34]

When the delegation of village headmen and deputies reached the Chigasaki (茅ヶ崎) area on 9.1.Hôei 5 (31.1.1708), they were met by the daimyo's representative Takatsuki Kansuke who was coming from Edo. The two Odawara officials Itô Bunnai and Hironaka Iemon who had previously gone up to Edo also appeared on behalf of the domain, as well as Ôtsu Zensaemon and an additional official from Odawara, Sugiyama Kouemon (杉山小右衛門). Takatsuki Kansuke informed the village representatives that their lord was well aware of their plight and had ordered relief rice to be distributed, but details would only be forthcoming once they had returned to Odawara. To placate the representatives, he also explained that previously the official Yanagida Kyûzaemon had not correctly transmitted the lord's intentions and would no longer be sent to the area.[35]

After some consultation the representatives indicated that while they themselves fully agreed to these terms, if they now returned to Odawara, the throng of several thousand farmers would once again set out for Edo. The government would take this to be an uprising and this would greatly damage the reputation of the domain. They therefore suggested that they would proceed with Takatsuki Kansuke to Edo to listen to the lord's proposal there, unless the officials preferred to inform them on the spot. The officials had no alternative but to choose the latter option. They said that their lord would give the farmers all he had in store, namely 20,000 bales (*hyô* 俵) of relief rice, even though this meant curbing his own lifestyle. With regard to removal of the tephra, however, all he could do was to express his deep empathy for the hard task facing the farmers.

The representatives expressed their gratitude for the relief rice, but argued that if they returned without any positive news about the removal of the fall-out, the farmers would set out for Edo again, and they hence decided to continue their journey. The domain administration at Edo was obviously alarmed by the approach of the village representatives, for a further official, Kanô Gôsuke (加納郷助), was sent to meet them when they reached Fujisawa. To back him up, Ôtsu Zensaemon and Sugiyama Kouemon arrived from Odawara. Kanô Gôsuke now informed the village representatives in the name of the domain lord that they would be receiving help to remove the fall-out in the amount of 27,000 gold coins (*ryô* 両), and if some villages still felt the need for more assistance, they could tender requests at the local domain offices. With these promises the village representatives finally returned to Odawara.[36]

On hearing the news that the village representatives were heading for Odawara, all the farmers assembled for their arrival on 10.1.Hôei 5 (1.2.1708). On the next day the representatives handed to Takatsuki Kansuke and Kanô Gôsuke—the two officials who had been sent from Edo—a long document recording the whole process of negotiations with the domain, including the reasons for their actions.[37] They likely thought the episode concluded, but that was not the case.

The following day, the village representatives were called to the domain office at Odawara. Here they faced an assembly of senior officials and were told that they had misunderstood the message given to them at Fujisawa. The amount of 27,000 *ryô* for the removal of the fall-out was simply an estimate and not a sum to be paid out by the domain. On hearing this, the throng of several thousand farmers assembled at Odawara began to riot. To diffuse the situation, the senior magistrate Kubota Jôemon (久保田丈衛門) now suggested that the village representatives journey to Edo, and set out ahead with Ôtsu Zensaemon.[38]

The village representatives left again for Edo on 13.1.Hôei 5 (4.2.1708), and when they approached Shinagawa on the very outskirts of the city the next day they were called into the presence of Kubota Jôemon and Ôtsu Zensaemon, as well as the official Kumamoto Sukedayû (熊本助太夫) who had come down from the Edo office. The latter explained that however much the farmers petitioned and complained, with the very large area of some 550,000 *koku* (石, 1 *koku*=4.9629 bushels) affected by the eruption, the domain did not have the capacity to deal with the disaster by itself, and bakufu help was being sought. He was going ahead to Edo to speak to a number of high-ranking government officials about the matter, and requested that the representatives delay their journey by a day.

Satisfied for the moment, the representatives stayed at Shinagawa and in the evening were again called to a meeting with domain officials. They were addressed by Kanô Gôsuke. That morning he had been called to a meeting with the domain lord and had briefed him on the situation. The lord was extremely hurt and angry that even though he had expressed his deep-felt sympathy with the plight of the farmers, they were marching to Edo in protest, and he had suggested that they all be thrown into prison. But then the lord had to depart for Edo castle and when he returned from the seat of government his mood had altogether changed. He considered it pitiful that the plight of the farmers was such that they saw no other way out but marching to Edo. The 20,000 bales of relief rice would be sent, as well as extra rations for those starving. To pay for the removal of ashes, he would even sell his most treasured sword, forgo his evening meals, and if these funds were still not sufficient, would personally petition the shogun on behalf of the farmers. He swore to all this on his samurai honor and calling to witness a number of deities. These dramatic pledges sufficiently reassured the village representatives that they returned to Odawara the next day.[39]

An interesting point in this process is that the sentiments of the domain lord, the senior councilor Ôkubo Tadamasu, were very different before and after his visit to Edo castle, where most likely he attended upon the shogun. No doubt the shogun had reminded him that the people were "the foundation of the state" and that they had to be governed "so that they do not suffer from hunger cold…" as he had proclaimed shortly after his accession.[40] Two days later the government issued a statement complaining that no progress had been made on removing the heavy fall-out in the provinces of Musashi (武蔵), Sagami (相模) and Suruga (駿河), indicating that the shogun, or those responsible for enforcing his policies, were getting concerned about the management of the crisis. The order urged local officials (*jitô* 地頭) to take action to have the fields cleared for spring planting. While the farmers should be urged to do the work themselves, help should be provided where their labors alone were inadequate. Care should be taken to prevent famine. Further details were to be obtained from the finance magistrate, Ogiwara Shigehide (荻原茂秀, 1658–1713).[41]

This was an obvious interference in affairs over which domains typically exercised autonomous control. It foreshadowed the bakufu's order made public early the next month, namely that the shogun's administration would take direct control of the worst-hit areas in the provinces of Musashi, Sagami, and Suruga, and the daimyo and lesser landholders affected would be assigned landholdings in different areas.[42] But this order did not release the Odawara domain from providing the promised relief rations to the villagers. The domain tried to retract the promise to pay the extra rations to starving people, and there were questions about how rations should be divided among the villages, but eventually the promised relief was forthcoming and distributed in the first half of the intercalary (urû 閏) first month of Hôei 5.[43]

Relief under Bakufu Administration

On 18.urû1.Hôei 5 (9.3.1708), the village representatives were called to the village of Sakawa (酒匂, now part of Odawara) and were informed that until the land was cleared of the tephra and returned to its original state, it would be under the administration of the bakufu's Kantô (関東) magistrate, Ina Tadanobu (伊奈半左衛門忠順, 1672/1673?–1712), whose office would be located at Sakawa. The order for the jurisdictional change was issued in the name of the bakufu's finance magistrate Ogiwara Shigehide.[44] Like Shigehide, Tadanobu had risen to importance under the fifth shogun. The two men co-operated in the restoration of the area, exercising considerable authority, but both would come to fall not long after the death of the shogun.[45]

The areas affected by the fall-out required not only large expenditure to return the fields to their previous state; they also were unable to pay regular taxes until the task was completed. This state of affairs imposed a financial burden too heavy to bear for the daimyo domains or *hatamoto* (旗本, bakufu retainers) to whom the villages had been assigned as fiefs. Even the bakufu considered the extent of the damage too great to finance from its own budget and in an unprecedented measure decided to ask the whole country to contribute to the restoration of the affected areas. A flat tax of two golden *ryô* per 100 *koku* of assessed rice yield was to be contributed by landholders. Only the areas hit by the eruption, temple lands, and holders of fiefs of less than fifty *koku* were exempt.[46] Those with landholdings of over 10,000 *koku*, namely the daimyo, had till the third month to pay their due, while lesser landholders, the *hatamoto*, were given until the sixth month to provide the funds.

Payments to villages were forthcoming without significant delay. The 20,000 *hyô* promised by the Odawara domain was distributed according to the depth of the deposit. The record of an unidentified village in the Sakawa area notes that the settlement was allotted 116 *hyô*, half of which they received in rice and half in cash.[47] Following that were distributions of funds which the local record clearly identifies as the shogun's tax on the whole country. The village in question received forty-nine *ryô* (両) two *bu* (分) in gold coins as well as one *kan* (貫) 115 *mon* (文) in smaller coins twice that year, once in the third and once in the twelfth month. Further, a rice ration of one *bu* and 530 *mon* was distributed for fields that were being cleared of debris, and three *bu* 706 *mon* for every two horses that had to be fed.[48]

In other villages where houses were almost totally destroyed, money was paid for rebuilding. One of the best-known examples is the village of Subashiri close to the source of eruption, which an early government report had described as being totally buried under the fall-out. Situated near the Kagosaka mountain pass (籠坂峠) that linked the politically important domains of Suruga and Kai (甲斐), the village was charged with the task of providing transport and accommodation for travelers. Hence the government considered the rebuilding of the village a priority and paid owners according to the size of the buildings lost: one *ryô* for every *tsubo* (坪, 1 *tsubo*=3.31 square meter) for buildings burnt down, and two *bu* for every *tsubo* for buildings crushed by the fallout. With thirty-seven buildings burnt and thirty-five buildings crushed, the village received 1,811 *ryô*. Here the record details the location of houses, the width of the frontage and their depth to the back, the size of the floor area, and the name of the owner, as Figure 7.4 shows.

The village also provided accommodation and guides for worshippers wishing to climb Mount Fuji and was economically fortunate in as much as the numbers

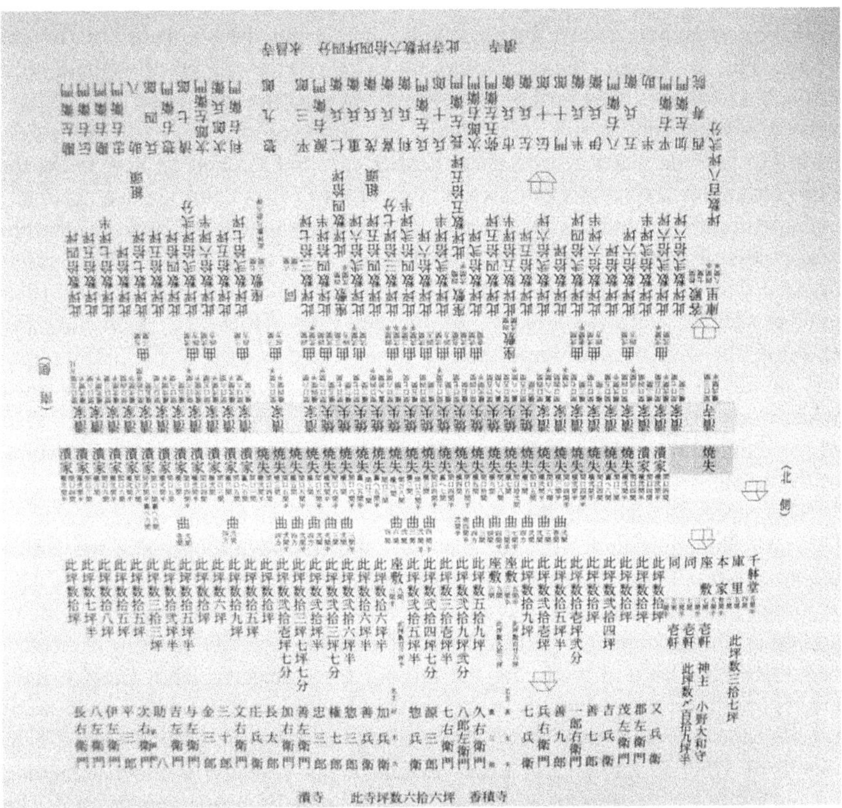

Figure 7.4 Residences in Subashiri Village showing burnt (pink) and crushed (yellow) houses after the eruption, detailing the location, size, and owner (source: National Museum of Japanese History (ed.), *Documenting Disaster: Natural Disasters in Japanese History, 1703–2003* (2003), p. 66).

of visitors greatly increased after the eruption. The coins the pious threw into Fuji's crater were collected and divided among the villagers and the local temple. The record shows a considerable increase in funds as the number of visitors rose, who also might have been making higher donations than before in an effort to placate the irate mountain.[49]

Much as in our own era when the contaminated soil was raked together and stored after the Fukushima disaster, the farmers attempted to clear their fields by piling up the tephra in certain locations. However, the difference was that they had no plastic bags to secure the tephra in one location and that the region was not flat, like that around Fukushima, but sloping. When heavy rains began in the spring, the accumulated tephra washed down the slopes, clogged river beds, and caused flooding.

Repairing the Sakawa River

One area prone to flooding even before the eruption was along the Sakawa River (酒匂川), stretching some forty-six kilometers from the southeastern foot of Mount Fuji to Sagami Bay at Odawara. The river's natural flow had been through a low-lying area further to the west, but after the area came into the possession of the Ôkubo family, they altered the course of the river early in the Edo period to provide water also to areas further east and to open up new fields for rice cultivation. To redirect the water, the Ôguchi (大口) and Garase (岩流瀬) embankments were built. These held during normal rainfall, but broke when rain was excessive, creating extensive flood damage. Prior to the Fuji eruption, major damage was on record already for 1670 and 1671 (Kanbun 寛文10–11), 1688 (Jôkyô 貞享5), 1695 (Genroku 元禄 8), and 1704 (Hôei 1).[50] Soon after the eruption, the river became impossible to traverse and as early as the first intercalary month of Hôei 5, the bakufu responded by charging five daimyo to assist with river works, exercising one of the bakufu's established prerogatives.[51] One of the daimyo was Ikeda of Okayama, where a record has remained with some of the details.

The representatives of the daimyo were called in by the aforementioned Ogiwara Shigehide and informed that they did not have to provide the actual workers, but only supervisors. The work itself was tendered to contractors who in turned hired local workers paying them one *momme* (匁) five *bu* (分), with an additional three *momme* to village heads. The work included cleaning the riverbeds of tributaries and removing the fall-out from adjacent land. Despite their efforts, as soon as sediments were removed from the riverbed, more were washed down from higher areas. Work extended from the second to the sixth month of Hôei 5, but a local observer noted that it ended without achieving much, while the contractor was said to have made a handsome profit.[52] The bakufu, on the other hand, believed that the task had been successfully completed and on 28.6.Hôei 5 recommended and rewarded the five daimyo to whom the task had been assigned, as well as Ogiwara Shigehide and Ina Tadanobu.[53] However, local records document that even before that date, on 22.6.Hôei 5, strong winds and heavy rains caused parts of the river embankment to break, flooding the surrounding villages and washing away houses. Further breaches in the embankment occurred early the next month, on 2.7.Hôei 5, so that finally six villages, Madarame (斑目), Sendoshima (千津島), Okano (岡野), Mamashita (儘下), Takematsu (竹松), and Wadakawahara (和田川原), were totally destroyed. The farmers petitioned Edo early the next year, and magistrate Ina Tadanobu paid out relief seven times during the course of that year with one unidentified village receiving a total of twelve *ryô* three *bu* plus 400 *mon*. To relocate the inhabitants of the six villages washed away, permission was given to cut down domain and bakufu forests to provide new land for the displaced inhabitants. In the end it would take nearly one hundred years until the greater part of the flooding problem was solved, but even now the river floods the Ashigara plain in periods of heavy rain.[54]

No bakufu record exists of the total damage caused by the eruption or funds paid out, and scholars have attempted to reconstruct both by using local records. One recent publication comes to the conclusion that there were four locations where houses were totally burnt or crushed by the pyroclastic fallout, including the village of Subashiri discussed above. Large mudflows caused destruction in twelve areas, including the total destruction of six villages mentioned earlier. In ten localities, including Hiratsuka, Fujisawa, and Yokohama, tephra clogged up irrigation channels and caused floods.[55]

What distinguishes the restoration of the stricken areas after the eruption of Mount Fuji in 1707 relative to recovery from the eruption of Mount Asama in 1783 is that it proceeded without famines or riots. This was achieved by placing the areas worst hit under direct shogunal control and making the whole country responsible for the restoration by levying a tax on all fiefs. After Mount Asama's eruption, the government refused to take on responsibility for the stricken areas, and this resulted in large-scale destruction (*uchi kowashi* 打毀) by riots of desperate farmers.[56]

In a system where the central government was expected to finance from the income of its own domain the administration of the whole country, the new tax of two *ryô* per one hundred *koku* of income on domains was an unprecedented interference in the finance of daimyo and lesser fief-holders. Needless to say, the measure was extremely unpopular with those who were asked to pay. After the death of the fifth shogun, this tax was cited as yet another example of misgovernment initiated during his administration, and still appears in history books today as such. While historians feel they cannot censor the shogun for wishing to feed and assist starving villagers, it has been suggested that the greater part of the funds collected were embezzled either by the bakufu or the unpopular finance magistrate Ogiwara Shigehide who was considered the architect of the plan. Adding up extant records, scholars come to the conclusion that of the 488,770 *ryô* collected, only 60,000 *ryô* were actually used for the relief program.[57] But even Arai Hakuseki (新井白石, 1657–1725), the Confucian advisor to Tsunayoshi's successor, Tokugawa Ienobu (徳川家宣, 1662–1712), who was highly critical of Tsunayoshi's government, noted that by the time the fifth shogun passed away not long after the New Year's celebrations of Hôei 6 (1709), 160,000 *ryô* had been spent on relief.[58] Since the stricken areas required support on a long-term basis, it is perhaps natural that just over a year after the eruption only a part of the funds had been used.

With Tsunayoshi's death, the bakufu was faced with a variety of unusual expense. There were the funeral expenses and the construction of a mausoleum which Hakuseki estimated at 200,000 *ryô*. Further, the imperial palace at Kyoto had burned down and the reconstruction was completed at some 700,000–800,000 *ryô* in the second half of Hôei 6 (1709). However, what even worsened the financial situation was the sixth shogun's request for the construction of a new residence at the cost of some 700,000 *ryô*.[59] In contrast, the fifth shogun on his accession had moved into the quarters of his predecessor, requesting only a new toilet, but even then offered to fast while the toilet was renewed so as not to incur the expense of building an additional facility.[60] With the succession of the sixth shogun, the relief program lost the importance the fifth shogun had assigned to it, and aid to the stricken areas was reduced and then mostly ceased.

Judging that agricultural areas with pyroclastic deposits of over some ninety centimeters (three *shaku*) would be difficult to restore, the bakufu was providing the population that remained in these areas with an emergency ration the equivalent of one *gô* (0.18 liters) of rice per person. The record shows, for instance, that this was paid to thirty-nine villages in Suntô gun (駿東郡) of Suruga province. However, these payments stopped abruptly in 2.Hôei 6, one month after the death of the fifth shogun. The village representatives went to Edo to request relief from Ina Tadanobu, who sent his officials for inspection in the third and fourth month, with the result that people considered near starvation were at least given five *mon* per person for five days.

Areas with less than 90 cm of fallout were paid rations to enable them to remove the deposits themselves on a pro rata basis depending on the depth of fallout, varying between one gold *bu* to 300 *mon* per *tan* (担, 1 *tan*=993 square meter) of land. The payment was, however, only made twice, in the third and eleventh month of Hôei 5 (1708), and discontinued after the death of the shogun. After petitions by the villagers, Ina Tadanobu accompanied by a delegation of officials inspected the area in the fifth and sixth month of Hôei 6 (1709). This at least led the bakufu in the eight month to again request domains to execute river works, providing employment for villagers as laborers. One village document records that on another occasion, Tadanobu personally took a village delegation to the residence of Ogiwara Shigehide, and managed to obtain increased wages for farmers employed in river works.[61] However, after the middle of Hôei 7 (1710), some one and a half years after the death of the fifth shogun, the bakufu stopped ordering domains to carry out river works, abolishing this policy much resented by the daimyo. The effect was that not only sediments were no longer cleared and embankments repaired, but also that the villagers lost the opportunity to earn wages as laborers.[62]

During this time we can also detect a change in officials responsible for the work. While previously the main responsibility for the recovery of the stricken areas lay with Ogiwara Shigehide and Ina Tadanobu, their duties started to be transferred to new men appointed by the new administration. Shigehide and Tadanobu were still officially rewarded for their work in the stricken areas, but now they were merely names in a list of many, and their rewards were often less than those of others.[63]

Eventually both Shigehide and Tadanobu's responsibilities came to an end. Even though Shigehide and his financial policies were strongly opposed by Arai Hakuseki, the Confucian advisor of the sixth shogun Ienobu, the latter continued to rely on Shigehide's financial expertise. Only when Ienobu lay on his death-bed, in the ninth month of Shôtoku (正德) 2 (1712), did Hakuseki succeed in bringing Shigehide down.[64] Already in the second month of that year Ina Tadanobu had passed away under mysterious circumstances. Tradition has it that unable to obtain funds to feed the people in his charge, Tadanobu opened government rice stores at Konyamachi (紺屋町) in Sunpu (駿府) and distributed 13.000 *koku* of rice to the starving farmers without government authorization. In atonement for this act of disobedience he is believed to have committed *seppuku*. He came to be venerated by the local population. The fact that a shrine honoring his memory was only erected in Keiô (慶應) 3 (1867), when the end of the Tokugawa bakufu was imminent, may be an indication that Tadanobu had fallen out of grace with the regime, a government that after the death of the fifth shogun no longer considered the farmers "the foundation of the state" to be governed "so that they do not suffer from hunger cold…"[65]

"People's power surged like the in-coming tide (民力さし潮のごとく)" Tokugawa Nariaki (徳川斉昭, 1800–1860), lord of Mito (水戸) wrote about the government of the fifth shogun as he reflected on the origins of popular uprisings and riots plaguing the Tokugawa government during his lifetime.[66] Over one hundred years later, the ruling elite were still resentful of Tsunayoshi's policies supporting the farmers and were holding him responsible for the gradual erosion of their authority. For the commoners, however, these policies set a precedent of how the government should respond in the case of natural disaster. When the bakufu failed to do so, the resulting hardship led to social upheaval and widespread disturbances, hastening the fall of the Tokugawa regime.

Notes

1 Koyama Masato 小山真人 and Maejima Yoshinori 前嶋美紀, English titles supplied for Japanese publication: "Estimate of the Height of Eruption Column of the 1707 Hoei Eruption of Fuji Volcano, Japan, Based on Historical Documents and CG Scenery Construction 古記録と景観再現CGにもとづく1707 年富士山宝永噴火の噴煙柱高度推定," in *Collection of Lectures at Japanese Association of Volcano* [日本火山学会講演予稿集], (Volcanological Society of Japan, 2009), p. 8. The altitude of 20 kilometers has been arrived at by examining contemporaneous reports and calculating the distance from the eruption of the observer.

2 Miyaji Naomichi, Kan'no Ayumi, Kanamaru Tatsuo, and Mannen Kazutaka, "High Resolution Reconstruction of the Hoei Eruption (AD 1707) of Fuji Volcano, Japan," The Smithsonian/NASA Astrophysics Data System, DOI:10.1016/j.jvolgeores.2011.06.013. Accessed on 1 October 2013.

3 The Hôei eruption receives the lowest ranking, namely one out of eleven in a work showing the frequency with which historical events are mentioned in the high school text *Nihon Rekishi B*. See Zenkoku rekishi kyôiku kenkyû kyôgikai 全国歴史教育研究協議会 (National History Education and Research Council) (ed.), *Nihon rekishi B yôgo shû* 日本歴史B用語集 [*Terminology of Japanese History B*], (Toyko: Yamakawa Shuppansha 山川出版社, 2004), p. 165.

4 For details on the fifth shogun's bad reputation see B. M. Bodart-Bailey, *The Dog Shogun: The Personality and Policies of Tokugawa Tsunayoshi* (Honolulu: University of Hawai'i Press, 2006), especially pp. 1–9.

5 See for instance Koyama Masato 小山真人, *Fuji funka no haza-do mappu: Hôei funka no 16 nichi kan* 富士噴火のハザードマップ一宝永噴火の16日間一[*Hazard Map of the Eruption of Mt. Fuji—the Sixteen Days of the Hôei Eruption*], (Tokyo: Kokin shoin 古今書院, 2009).

6 Nakajima Shigehiko 中島林彦, "Ukabiagaru su-pa-saikuru 浮かび上がるスーパーサイクル (The Emerging Super-Cycle)," in *Asahi saiensu* 朝日サイエンス: *Shinsai to genpatsu* 震災と原発 [*Asahi Science: Earthquakes and Nuclear Energy*], No. 183 (16.2.2012): 25–48.

7 H. Tsuya, "Geological and Petrological Studies of Volcano Fuji, V. 5. On the 1707 Eruption of Volcano Fuji," *Bull. Earthquake Research Institute*, 33 (1955): 341, 343.

8 Bodart-Bailey, *The Dog Shogun*, pp. 255–260.

9 Miyaji Noaomichi 宮地直道 and Koyama Masato 小山真人, English title supplied for Japanese publication: "Recent Studies on the 1707 (Hoei) Eruption of Fuji 富士火山1707 (宝永噴火) についての最近の研究成果," in Aramaki Shigeo 荒牧重雄 et al. (eds.), *Fuji Kazan* [富士火山], (2007), p. 339. Tsuji Yoshinobu つじよしのぶ, *Fujisan no funka: manyôshû kara gendai made* 富士山の噴火：万葉集から現代まで [*Mount Fuji's Eruptions: From the Manyôshû to the Present Day*], (Tokyo: Chikuchi Shokan 築地書館, 1992), p. 183, put the magnitude at 8.4. For a description in English see Bodart-Bailey, *The Dog Shogun*, p. 265.

10 Inscription on a contemporary illustration, see National Museum of Japanese History 国立歴史民俗博物館 (ed.), *Documenting Disaster: Natural Disasters in Japanese History, 1703–2003* [ドキュメント災害史 1703－2003] (2003), p. 60, fig. 2.6.

11 National Museum of Japanese History (ed.), *Documenting Disaster*, p. 60.

12 For the shape and composition of the craters and Mount Hôei see Tsuya, "Geological and Petrological Studies," pp. 355–359.

13 Koyama Masato, *Hazard Map of the Eruption of Mt. Fuji*, p. 5.

14 Ibid.

15 Tsuya, "Geological and Petrological Studies," p. 381.

16 Gojiin Ryûkô Sôjô 護寺院隆晃僧正, *Ryûkô sôjô nikki* 隆晃僧正日記 [*Diary of Ryûkô Sôjô*], in Nagashima Fukutarô 永島福太郎 and Hayashi Ryôshô 林亮勝, (eds.), 3 vols, Shiryô Sanshû 資料纂集 [*Collection of Documents*], (Zoku Gunsho Ruijû Kanseikai 続群書類従完成会, Tokyo, 1970), Vol. 3, p. 168, entry 23.11.Hôei 4. Ryûkô notes that doors and shoji were rattling from the early morning, but Koyama, comparing this with other records and the time of the explosive eruptions points out that "early morning" must be a mistake and that the rattling would have started around 10 a.m. (see Koyama, *Hazard Map of the Eruption of Mt. Fuji*, p. 59).

17 Miyaji Naomichi and Koyama Masato, "Recent Studies of the 1707 (Hoei) Eruption of Fuji Volcano," p. 339.

18 *Ryûkô sôjô nikki*, Vol. 3, p. 168, entry 23.11.Hôei 4.

19 Ui Tadahide 宇井忠英 et al., English title supplied for Japanese publication: "December 16, 1707 Ash Discharged from Mount Fuji: Fallout Archived in Tokyo 江戸市内に降下し保存されていた富士宝永噴火初日の火山灰," *Kazan*火山, 47:3 (2002): 87–88. Koyama Masato, *Hazard Map of the Eruption of Mt. Fuji*, pp. 75–76.

20 Ui et al., "December 16, 1707 Ash Discharged from Mount Fuji," pp. 90–93.

21 *Ryûkô sôjô nikki*, Vol. 3, pp. 168–169, entry 25.11.Hôei 4. Ms, *Bunrosô* 文露叢 [*Collection of Public Writings*], collection of Kokuritsu Kokkai Toshokan 国立国会図書館 (National Diet Library), Tokyo, entry 22–23.11.Hôei 4.

22 Kuroita Katsumi 黒板勝美 and Kokushi Taikei Henshûkai 国史大系編集会 (eds.), *Tokugawa Jikki* 徳川実記 [*Authentic Records of Tokugawa*], (Tokyo: Yoshikawa Kôbunkan 吉川弘文館, reprint 1976), Vol. 6, p. 675, entry 25.22.Hôei 4. Koyama, *Hazard Map of the Eruption of Mt. Fuji*, pp. 14–15. The position of *okachi metsuke* (御徒目付) is below that of *metsuke*. The *okachi metsuke* have been described as "detectives" and on twenty-four-hour duty at Edo castle would have been ready to set off immediately. See Sasama Yoshihiko 笹間良彦, *Edo bakufu yakushoku shûsei* 江戸幕府役職集成 [*Compilation of Offices of the Edo Bakufu*] (Tokyo: Yûzankaku 雄山閣, 1965), pp. 357–358. Kawaguchi Kenji 川口謙二 et al., *Edo jidai yakushoku jiten* 江戸時代役職事典 [*Dictionary of Offices of the Edo Period*], (reprint 1992), p. 26.

23 Koyama, *Hazard Map of the Eruption of Mt. Fuji*, pp. 15–16 citing *Odawara shishi* 小田原市史 [*History of Odawara City*].

24 Clive Oppenheimer, *Eruptions that Shook the World*, (Cambridge: Cambridge University Press, 2011), pp. 30–31.

25 Explanations (kaidai 解題) of Ôtomo Kazuo 大友一雄 of "Fujisan sunafuri sogan kiroku 富士山砂降り訴願記録 (Record of Petitions Concerning Sand Erupting from Mount Fuji, Sagami)," by Suzuki Rizaemon 鈴木利左衛門 in *Nihon nôsho zenshû* 日本農書全集 [*Complete Edition of Japanese Agricultural Work*], Vol. 66, (Tokyo: Nôsan Gyoson Bunka Kyôkai 農山漁村文化協会, 1994), pp. 72–73.

26 For a discussion of this subject, see Bodart-Bailey, *The Dog Shogun*, especially pp. 231–232.

27 For details see ibid., especially pp. 2–9.

28 Suzuki Rizaemon 鈴木利左衛門, "Fujisan sunafuri sogan kiroku (Sagami) 富士山砂降り訴願記録(相模) (Record of Petitions Concerning Sand Erupting from Mount Fuji, Sagami)," in *Nihon nôsho zenshû* 日本農書全集 [*Complete Edition of Japanese Agricultural Works*], Vol. 66, (Tokyo: Nôsan Gyoson Bunka Kyôkai 農山漁村文化協会, 1994), pp. 34–36.

29 Suzuki Rizaemon, pp. 36–38. The document does not specify the title of Yanagida Kyûzaemon, but only states that he was an official (御役人).

30 Ibid., pp. 38–39.

31 Ibid., pp. 38–39.

32 Ibid., pp. 41–42.

33 Ibid., pp. 42–43.

34 Ibid, pp. 43–44. The position of men described as "officials" does not appear in the document and could not be determined.

35 Ibid., pp. 44–45.

36 Ibid., pp. 45–47.

37 Ibid., pp. 47–53.

38 Ibid., pp. 54–55.

39 Ibid., pp. 56–57.

40 Bodart-Bailey, *The Dog Shogun*, p. 91.

41 *Tokugawa Jikki*, Vol. 6, p. 679, 16.1.Hôei 5 reproduces the order except for the reference to Ogiwara Shigehide. The complete order is contained in *Bunrosô* ms.; Naikaku bunko, Vol. 5, 21.1.Hôei 5 as well as Takayanagi Shinzô 高柳信三 and Ishii Ryôsuke 石井良助, *Ofuregaki kanpô shûsei* 御触書寛保修成 [*Collection of Government Orders of the Kanpô Period*], (Tokyo: Iwanami Shoten 岩波書店, 1958), 25: 745–746.

42 *Tokugawa Jikki*, Vol. 6, p. 681, 3.urû1.Hôei 5. An intercalary month (*urû*) was inserted that year to adjust the lunar calendar to the solar cycle.

43 Suzuki Rizaemon, pp. 59–65.

44 Ibid., pp. 66–67.
45 For details see Bodart-Bailey, *The Dog Shogun*, pp. 271–273.
46 *Tokugawa Jikki*, Vol. 6, p. 682, 7.urû1.Hôei 5. One *koku* is approximately five bushels. The bakufu had until this point financed all its expenditure from its own revenues.
47 Judging from the facts contained in the document, the author is believed to have lived in one of the villages of Madarame (斑目), Sendoshima (千津島), Okano (岡野), Mamashita (儘下), Takematsu (竹松), or Wadakawahara (和田川原). Izumi Masahiro 泉雅博, Introduction (kaidai 解題) to "Fujisan yakidashi suna ishi furi no koto 富士山焼き出し砂石降りのこと (Concerning the Eruption of Sand and Stones from Mount Fuji)," in *Nihon nôsho zenshû* 日本農書全集 [*Complete Edition of Japanese Agricultural Works*], Vol. 66, (Tokyo: Nôsan Gyoson Bunka Kyôkai 農山漁村文化協会, 1994), pp. 124–125.
48 Author unknown, "Fujisan yakidashi suna ishi furi no koto 富士山焼き出し砂石降りのこと(Concerning the Eruption of Sand and Stones from Mount Fuji)," in *Nihon nôsho zenshû* 日本農書全集 [*Complete Edition of Japanese Agricultural Works*], Vol. 66, (Tokyo: Nôsan Gyoson Bunka Kyôkai 農山漁村文化協会, 1994), pp. 107–108.
49 Takano Yoshihiko 高野義彦, *Genroku, Kyôhô no jidai* 元禄享保の時代 [*The Periods of Genroku and Kyôhô*], Vol. 13 of *Nihon no rekishi* 日本の歴史 [*History of Japan*], (Tokyo: Shûeisha 集英社, 1992), pp. 205–207. *Documenting Disaster: Natural Disasters in Japanese History, 1703–2003*, pp. 66–67. Matsuo Mieko 松尾美恵子, "Fujisan funka to Asamasan funka 富士山噴火と浅間山噴火 (The Eruption of Mount Fuji and the Eruption of Mount Asama)," in Ôishi Manabu 大石学 (ed.), *Kyôhô kaikaku to shakai henyô* 享保改革と社会変容 [*The Reforms of the Kyôhô Period and Social Change*], Vol. 16 of *Nihon no jidai shi* 日本の時代史 [*Era History of Japan*], (Shûeisha 集英社, Tokyo, 2003), pp. 149, 157.
50 Izumi Masahiro, Introduction to "Concerning the Eruption of Sand and Stones from Mount Fuji," pp. 127–129.
51 *Tokugawa jikki*, Vol. 6, p. 682, 9.urû1.Hôei 5.
52 "Concerning the Eruption of Sand and Stones from Mount Fuji," p. 108. Matsuo Mieko, "The Eruption of Mount Fuji and the Eruption of Mount Asama," pp. 156–157.
53 *Tokugawa jikki*, Vol. 6, p. 700, 28.6.Hôei 5.
54 "Concerning the Eruption of Sand and Stones from Mount Fuji," pp. 109–110. Inoue Kimio 井上公夫, English title supplied for Japanese publication: "Distribution of Sediment-related Disasters after the Hoei Eruption of Fuji Volcano in 1707, Based on Historical Documents 富士山宝永噴火 (1707) 後の長期間に及んだ土砂災害," in Aramaki Shigeo et al. (eds.), *Fuji kasan*, pp. 431–433. See the HP of Kanagawa Prefectural Government, Sakawagawa suikei Sakawagawa shinsui sôtei kuiki chizu 酒匂川水系酒匂川浸水想定区域図 for a map and description of present-day flooding. www.pref.kanagawa.jp/cnt/f3747/p11314.html, accessed on 1 October 2013.
55 Inoue Kimio, "Distribution of Sediment-related Disasters after the Hoei Eruption of Fuji Volcano in 1707," pp. 427–429.
56 Matuso Mieko, "The Eruption of Mount Fuji and the Eruption of Mount Asama," pp. 172, 174–175.
57 For instance Ôtomo Kazuo 大友一雄, *Fujisan funka go no fukkyu kōji*富士山噴火後の復旧工事 in *Nihon nôsho zenshû*, Vol. 66, p. 77; Takano Yoshihiko, *Genroku, Kyôhô no jidai*, p. 207.
58 Arai Hakuseki 新井白石, *Arai Hakuseki nikki* 新井白石日記 [The Diary of Arai Hakuseki], in Tôkyô Daigaku Shiryô Hensanjo 東京大学史料編纂所 (ed.), *Dai nihon ko kiroku* 大日本古記録 [*Ancient Documents of Japan*], (Tokyo: Iwanami Shoten 岩波書店, 1953), Vol. II, p. 83. *Oritaku shiba no ki* 折りたく柴の記, in *Nihon bungaku koten taikei* 日本文学古典大系 [*Major Works of Japanese Classical*

Literature], reprint 1967, Vol. 95, p. 233. Translated by Joyce Ackroyd as *Told Round a Brushwood Fire* (Tokyo: University of Tokyo Press, 1979), pp. 100–101.
59 *Oritaku shiba no ki*, Vol. 95, pp. 232, 238, 305. *Told Round a Brushwood Fire*, pp. 100, 104, 169.
60 Hotta Masatoshi 堀田正俊, *Yôgen roku* 颺言録 [*The Record Made Public*], in *Zoku zoku gunsho ruijû* 続々群書類従 1906–1909 [*Continued Collection of Books, 1906–1909*] Vol. 13, p. 33.
61 Matuso Mieko, "The Eruption of Mount Fuji and the Eruption of Mount Asama," pp. 157–158. Komiya Masaaki 古宮雅明, English title supplied for Japanese publication: "River Dredging in Mt. Fuji Houei Eruption Stricken Area, and the Countermeasure of Tokugawa Shogunate 富士山宝永噴火被災地の川普請と幕府の対応," *Bulletin of the Kanagawa Prefectural Museum, Cultural Sciences* [神奈川県立博物館研究報告－人文科学—] No. 35 (2009): 30–32.
62 Komiya Masaaki, "River Dredging," p. 33, p. 37.
63 Ibid., pp. 43–44.
64 For details see Bodart-Bailey, *The Dog Shogun*, pp. 272–273.
65 Honma Kiyotoshi 本間清利, *Kantô gundai: Ina shi no keifu* 関東郡代: 伊奈氏の系譜 [*Magistrates of the Kantô Area: The Genealogy of the Ina Family*], (Saitama Shinbunsha 埼玉新聞社, Saitama, 1983), p. 189. Arai shukueki to chiiki machizukuri kyôgikai 新井宿駅と地域まちづくり協議会, "Hôei funka to Ina Hanzaemon Tadanobu 宝永噴火と伊奈半左エ門忠順," available at: http://araijyuku.jp/history/edo/inatadanobu.pdf, accessed on 1 October 2013.
66 Tokugawa Nariaki 徳川斉昭, cited in Fujita Satoru 藤田覚, *Kindai no taidô* 近代の胎動 [*The Stirring of the Modern Age*], Vol. 17 of *Nihon no jidaishi* 日本の時代史 [*Era History of Japan*], (Tokyo: Yoshikawa kôbunkan 吉川弘文館, 2003), p. 67.

Bibliography

Manuscript

Bunrosô 文露叢 [*Collection of Public Writings*], Tokyo: Kokuritsu Kokkai Toshokan 国立国会図書館 (National Diet Library).

Published works

Arai Hakuseki 新井白石. *Arai Hakuseki nikki* 新井白石日記 [*The Diary of Arai Hakuseki*]. Tôkyô Daigaku Shiryô Hensanjo 東京大学史料編纂所 ed., *Dai nihon ko kiroku* 大日本古記録 [*Ancient Documents of Japan*]. Tokyo: Iwanami Shoten 岩波書店, 1953.

Arai Hakuseki 新井白石. *Oritaku shiba no ki*折りたく柴の記 [*Told Round a Brushwood Fire*] in *Nihon bungaku koten taikei*日本文学古典大系 [*Major Works of Japanese Classical Literature*], reprint 1967, Vol. 95, p. 233. Translated by Joyce Ackroyd as *Told Round a Brushwood Fire*, Tokyo: University of Tokyo Press, 1979.

Aramaki Shigeo 荒牧重雄 et al. (eds.). *Fuji Kazan* 富士火山 [*Fuji Volcano*]. Yamanashi: Yamanashi Institute of Environmental Sciences 山梨県環境科学研究所, 2007.

Author unknown, "Fujisan yakidashi suna ishi furi no koto 富士山焼き出し砂石降りのこと(Concerning the Eruption of Sand and Stones from Mount Fuji)," in *Nihon nôsho zenshû* 日本農書全集 [*Complete Edition of Japanese agricultural works*], Vol. 66, pp. 107–108.

Bodart-Bailey, Beatrice, M. *The Dog Shogun: The Personality and Policies of Tokugawa Tsunayoshi*, Honolulu: University of Hawai'i Press, 2006.

Gojiin Ryûkô Sôjô 護寺院隆晃僧正, *Ryûkô sôjô nikki* 隆晃僧正日記 [*Diary of Ryûkô Sôjô*], in Nagashima Fukutarô 永島福太郎 and Hayashi Ryôshô 林亮勝 (eds.), *Shiryô Sanshû* 資料纂集 [*Collection of Documents*], Tokyo: Zoku Gunsho Ruijû Kanseikai 続群書類従完成会, 1970, 3 Vols.

Honma Kiyotoshi 本間清利. *Kantô gundai; Ina shi no keifu* 関東郡代: 伊奈氏の系譜 [*Magistrates of the Kantô Area: The Genealogy of the Ina Family*]. Saitama: Saitama Shinbunsha 埼玉新聞社, 1983.

Hotta Masatoshi 堀田正俊. *Yôgen roku* 颺言録 [*The Record Made Public*], in *Zoku zoku gunsho ruijû* 続々群書類従, 1906–1909 [*Continued Collection of books, 1906–1909*], Vol. 13.

Inoue Kimio 井上公夫. English title supplied for Japanese publication: "Distribution of Sediment-related Disasters after the Hoei Eruption of Fuji Volcano in 1707, Based on Historical Documents 富士山宝永噴火(1707)後の長期間に及んだ土砂災害," in Aramaki Shigeo 荒牧重雄 et al. (eds.), *Fuji Kasan*富士火山 [*Fuji Volcano*]. Yamanashi: Yamanashi Institute of Environmental Sciences 山梨県環境科学研究所, 2007, pp. 431–433.

Izumi Masahiro 泉雅博. Introduction (解題) to "Fujisan yakidashi suna ishi furi no koto 富士山焼き出し砂石降りのこと(Concerning the Eruption of Sand and Stones from Mount Fuji)," in *Nihon nôsho zenshû* 日本農書全集 [*Complete Edition of Japanese Agricultural Works*], Vol. 66, Tokyo: Nôsan Gyoson Bunka Kyôkai 農山漁村文化協会, 1994.

Kawaguchi Kenji 川口謙二 et al. *Edo jidai yakushoku jiten* 江戸時代役職事典 [*Dictionary of Offices of the Edo Period*], reprint 1992.

Komiya Masaaki 古宮雅明. English title supplied for Japanese publication: "River Dredging in Mt. Fuji Houei Eruption Stricken Area, and the Countermeasure of Tokugawa Shogunate 富士山宝永噴火被災地の川普請と幕府の対応," 神奈川県立博物館研究報告－人文科学— [*Bulletin of the Kanagawa Prefectural Museum, Cultural Sciences*], No. 35 (2009): 25–72.

Koyama Masato 小山真人. *Fuji funka no haza-do mappu: Hôei funka no 16 nichi kan* 富士噴火のハザードマップ－宝永噴火の16日間－ [*Hazard Map of the Eruption of Mt. Fuji—the Sixteen Days of the Hôei Eruption*], Tokyo: Kokin shoin 古今書院, 2009.

Koyama Masato 小山真人 and Maejima Yoshinori 前嶋美紀. English title supplied for Japanese publication: "Estimate of the Height of Eruption Column of the 1707 Hoei Eruption of Fuji Volcano, Japan, Based on Historical Documents and CG Scenery Construction 古記録と景観再現CGにもとづく1707年富士山宝永噴火の噴煙柱高度推定," in *Nihon kazan gakkai kôen yokô shû* 日本火山学会講演予稿集 [*Collection of Lectures at Japanese Association of Volcano*], Volcanological Society of Japan, 2009.

Kuroita Katsumi 黒板勝美 and Kokushi Taikei Henshûkai 国史大系編集会 (eds.), *Tokugawa Jikki* 徳川実記 [*Authentic Records of Tokugawa*]. Tokyo: Yoshikawa Kôbunkan 吉川弘文館, reprint 1976.

Matsuo Mieko 松尾美恵子. "Fujisan funka to Asamasan funka 富士山噴火と浅間山噴火 (The Eruption of Mount Fuji and the Eruption of Mount Asama)," in Ôishi Manabu 大石学 (ed.), *Kyôhô kaikaku to shakai henyô* 享保改革と社会変容 [*The Reforms of the Kyôhô Period and Social Change*]. Vol. 16 of *Nihon no jidai shi*日本の時代史 [*Era History of Japan*], Tokyo: Yoshikawa Kôbunkan 吉川弘文館, 2003.

Miyaji Noaomichi 宮地直道 and Koyama Masato 小山真人. English title supplied for Japanese publication: "Recent Studies on the 1707 (Hoei) Eruption of Fuji 富士火山

1707 (宝永噴火) についての最近の研究成果," in Aramaki Shigeo 荒牧重雄 et al. (eds.), *Fuji Kazan* 富士火山 [*Fuji Volcano*], Yamanashi: Yamanashi-ken Kagaku Kankyô Kenkyûjo 山梨県環境科学研究所, 2007.

Nakajima Shigehiko 中島林彦. "Ukabiagaru su-pa-saikuru 浮かび上がるスーパーサイクル (The Emerging Super-cycle)," in *Asahi saiensu* 朝日サイエンス *Shinsai to genpatsu* 震災と原発 [*Asahi Science: Earthquakes and Nuclear Power*], No. 183 (16.2.2012): 25–37.

National Museum of Japanese History 国立歴史民俗博物館 (ed.). *Documenting Disaster: Natural Disasters in Japanese History, 1703–2003*) [ドキュメント災害史 1703－2003], 2003.

Nihon rekishi B yôgo shû 日本歴史B用語集 [*Usage of Terminology in Japanese History B*], Zenkoku rekishi kyôiku kenkyû kyôgikai 全国歴史教育研究協議会, ed., Tokyo, 2004.

Oppenheimer, Clive, *Eruptions that Shook the World*, Cambridge: Cambridge University Press, 2011.

Ôtomo Kazuo 大友一雄. Introduction (解題) to "Fujisan sunafuri sogan kiroku 富士山砂降り訴願記録 (Record of Petitions Concerning Sand Erupting from Mount Fuji) by Suzuki Rizaemon," in *Nihon nôsho zenshû* 日本農書全集 [*Complete Edition of Japanese Agricultural Works*], Vol. 66, Tokyo: Nôsan Gyoson Bunka Kyôkai 農山漁村文化協会, 1994.

Sasama Yoshihiko 笹間良彦. *Edo bakufu yakushoku shûsei* 江戸幕府役職集成 [*Compilation of Offices of the Edo Bakufu*], Tokyo: Yûzankaku 雄山閣, 1965.

Suzuki Rizaemon 鈴木利左衛門. "Fujisan sunafuri sogan kiroku (Sagami) 富士山砂降り訴願記録(相模) (Record of Petitions Concerning Sand Erupting from Mount Fuji, Sagami)," in *Nihon nôsho zenshû* 日本農書全集 [*Complete Edition of Japanese Agricultural Works*], Vol. 66, Tokyo: Nôsan Gyoson Bunka Kyôkai 農山漁村文化協会, 1994.

Takano Yoshihiko 高野義彦. *Genroku, Kyôhô no jidai* 元禄享保の時代 [*The Periods of Genroku and Kyôhô*], Vol. 13 of *Nihon no rekishi* 日本の歴史 [*History of Japan*], Tokyo: Shûeisha 集英社, 1992.

Takayanagi Shinzô 高柳信三 and Ishii Ryôsuke 石井良助. *Ofuregaki kanpô shûsei* 御触書寛保修成 [*Collection of Government Orders of the Kanpô Period*], Tokyo: Iwanami Shoten 岩波書店, 1958.

Tsuji Yoshinobu つじよしのぶ, *Fujisan no funka: manyôshû kara gendai made* 富士山の噴火：万葉集から現代まで [*Mount Fuji's Eruptions: From the Manyôshû to the Present Day*], Tokyo: Chikuchi Shokan 築地書館, 1992.

Tsuya, H. "Geological and Petrological Studies of Volcano Fuji, V. 5. On the 1707 Eruption of Volcano Fuji," *Bull. Earthquake Research Institute*, 33 (1955), pp. 341–383.

Ui Tadahide 宇井忠英 et al. English title supplied for Japanese publication: "December 16, 1707 Ash Discharged from Mount Fuji: Fallout and Archived in Tokyo 江戸市内に降下し保存されていた富士宝永噴火初日の火山灰," *Kazan* 火山, 47:3 (2002).

Internet

Arai shukueki to chiiki machizukuri kyôgikai 新井宿駅と地域まちづくり協議会, "Hôei funka to Ina Hanzaemon Tadanobu 宝永噴火と伊奈半左エ門忠順 (The Hôei Eruption and Ina Hanzaemon Tadanobu)," http://araijyuku.jp/history/edo/inatadanobu.pdf, accessed on 1 October 2013.

Kanagawa Prefectural Government HP, Sakawagawa suikei Sakawagawa shinsui sôtei kuiki chizu 酒匂川水系酒匂川浸水想定区域図, www.pref.kanagawa.jp/cnt/f3747/p11314.html, accessed on 1 October 2013.

Miyaji Naomichi, Kan'no Ayumi, Kanamaru Tatsuo, and Mannen Kazutaka, "High Resolution Reconstruction of the Hoei Eruption (AD 1707) of Fuji Volcano, Japan," The Smithsonian/NASA Astrophysics Data System, DOI: 10.1016/j.jvolgeores.2011.06.013, www.sciencedirect.com/science/article/pii/S0377027311001879 accessed on 1 October 2013.

8 Across the Tama

Bridges and roads, rivers and rocks

M. William Steele

Introduction

Bridges and highways are everywhere symbols of modern life. Made of cement and steel, they speed the movement of peoples and flow of goods. They conquer rivers and mountains; overcoming obstacles they bring people together. This chapter focuses on the construction of Rokugō (六郷) Bridge across the Tama (多摩), a short river (138 km) that flows into Tokyo Bay between Tokyo (東京) and Yokohama (横浜), and the construction of Japan's first modern highway—the Keihin (京浜) Highway (National Highway Number 1)—completed in 1926, linking these two great cities.[1] The construction of bridges and highways have allowed for the development of a transportation system that has undeniably given people speed and convenience, but at a great cost to the natural environment. These projects heralded a boom in ferro-concrete construction starting before the Great Kanto Earthquake of 1923 and lasting into the early 1930s. These modern engineering projects have left a mixed legacy to Japan today; on the one hand the promise of unrestricted mobility, but at the same time massive environmental damage caused a burgeoning cement industry and government policies favoring construction.

A turbulent history

One of the first Western language accounts of the Tama River is by Egebert Kaempher who ferried across the "gentle, deep river" in 1691 on his way to Edo (江戸).[2] If he had traveled a few years earlier, he could have walked across the great Rokugō Bridge into the world's largest city. The bridge was originally built in 1600, soon after Tokugawa Ieyasu (徳川家康, 1543–1616) set up headquarters in Edo. It was part of the Tokaidō (東海道) Highway that linked Osaka and Kyoto with Edo, but the 218 meter wooden bridge suffered repeated damage by floodwaters. It was repaired in 1612 and 1643, washed out and rebuilt in 1646, 1659, and again in 1670. Floods destroyed the bridge in 1671 and 1672; it was rebuilt but damaged and repaired in 1684; washed out in 1686, repaired, but destroyed again in 1688.[3] At this point the Tokugawa government decided against building a new bridge. For the next 200 years ferries carried all traffic across the river; there were 27 crossings, but the Tokaidō Crossing at Rokugō was the busiest (see Figure 8.1). On several occasions, such as the transportation of an elephant in 1729, on special order from Shogun Tokugawa Yoshimune (徳川宗吉, 1684–1751), and the arrival of the young Meiji (明治) Emperor in the autumn of 1868, a special pontoon bridge was devised, but otherwise the Tama River hindered overland transport in and out of Edo.[4]

Other rivers on the Tokaidō were similarly un-bridged; sometimes the reason given was strategic, but often the causes were physical and social. Many of Japan's rivers were short and swift and flooded easily, especially in late summer and early autumn when rain was plentiful. During the Edo period, moreover, a spurt in agricultural development to support a growing population encouraged deforestation and the conversion of floodplains to farmland and villages. The result was an increase in the scale and incidence of floods. According to the *History of the Tama River* [多摩川誌], which lists all major incidents of flood damage along the Tama River, floods occurred on an average of every six years in the period between 1600 and 1700, and once every three years between 1700 and 1800.[5] The Tokugawa regime attempted various means to control the river. Special dykes, levees and embankments were built, channels were widened, and waters diverted. Specialists at water management, such as Tanaka Kyūgu (田中丘隅, 1662–1729) and later Kawasaki Heiyuemon (川崎平右衛門, 1694–1767) were invited to apply their talents on the Tama and other nearby rivers. They dug channels and constructed protrusions and other stone works at strategic places. However, flood damage to crops, agricultural communities and to bridgework only increased. In the first half of the nineteenth century, between 1800 and 1868, the river flooded on 30 occasions and with increasing ferocity, destroying any hope that the river could be bridged with existing technology.[6]

Bridging the Tama

After the Meiji Restoration of 1868, policies of rapid Westernization and industrialization placed new demands on the Tama and other river systems close to Tokyo. Flooding continued with even more frequency. In the 32 years between 1868 and 1900, flood years counted for 22. The Meiji government employed Western specialists to help solve the problem, but with little success.[7] A British engineer, Edmund Morel (1840–1871), oversaw the construction of a wooden railroad bridge across the Tama near the Rokugō Crossing as part of Japan's first railroad, 29 kilometers of track between Shinbashi (新橋) in Tokyo and Yokohama.[8] Construction of the 624 meter bridge was completed at the end of 1871, well in time for the opening of the new railroad on October 14, 1872. The bridge used stone abutments and was constructed with native timber by Japanese carpenters under Morel's supervision. Tracks were laid across board planks.

The railroad bridge became an immediate showcase for Japan's modernity (Figure 8.2); nonetheless, the bridge was condemned in 1875, after only three years of service.

Figure 8.1 A daimyo procession crosses the Tama River on the Tokaido post road at Kawasaki. The Ferry Crossing at Kawasaki 東海道川崎, Hiroshige II 重二代目, 1863 (source: Author's collection).

Figure 8.2 The Rokugō Railroad Bridge with the Rokugō Ferry Crossing upstream (Shōsai Ikkei 昇斎一景, *Rokugō riku jōkisha tetsudō no zenzu* 六郷陸蒸気車鉄道之全図, 1871) (source: Author's collection).

Richard Vicars Boyle (1822–1908), who succeeded Morel as Engineer-in-Chief of the Japanese Railways, ordered the renewal of all bridges and culverts with iron superstructures and permanent foundations and piers.[9] Boyle's new Rokugō River Bridge went into service in November 1877. This time no compromises were made. All ironwork was imported from England and the alignment and brick and cement work connected with the piers was carefully done in order to strengthen the bridge against flooding. According to the October 6, 1877 issue of the *Japan Weekly Mail*:

> When this new bridge is completed, it will be the longest in Japan, and in strength and general permanence of structure, may be almost termed everlasting, in which respect, it is unsurpassed by any bridge of its class in the world.[10]

In response to the increased volume of traffic between Yokohama and Tokyo, a private individual, Suzuki Sanai (鈴木左内, 1839–1909) the village headman of Yahatatsuka (八幡塚) Village, located on the Tokyo side of the Rokugō Crossing, applied for and received permission to construct and operate a toll bridge across the Tama to carry non-rail traffic, much to the consternation of the boatmen on the Kawasaki (川崎) side of the river who had a monopoly with the ferry business.[11] Completed in 1874, the wooden girder bridge was 109 meters in length and five meters above the water. It was quickly put to the test; heavy rain and wind destroyed it in 1875. The bridge was rebuilt, but suffered damage again in 1876 and 1877. The flood of September 16, 1878 carried away Sanai's toll bridge and burst the banks of the river at several points above and below the crossing, causing considerable loss of life and destruction of property. The newly built iron railroad bridge, located about 200 meters upstream, escaped damage. At this point Sainai gave up, allowing the boatmen and their flat-bottomed boats to resume their work unchallenged.

Later, in 1882, the mayor of Kawasaki and the chief of Yahatatsuka Village agreed to build a new toll bridge across the Tama at the Rokugō Crossing. They composed a joint petition addressed to the governor of Tokyo, Matsuda Michiyuki (松田道之, 1839–1882), and governor of Kanagawa Prefecture, Oki Morikata (沖守固, 1841–1912), seeking to build a 102 meter long and 5.5 meter wide wooden structure, to be called the Rokugō Bridge. It would be operated by a private company, the Rokugō Bridge Union (Rokugō-hashi kumiai 六郷橋組合), which promised to supplement rather than replace existing ferry services.[12]

The new wooden bridge and its successors met a similar fate, making ferry service the only dependable means of crossing the river for those who could not afford the train.[13] In 1906, the much-maligned bridge was placed under the administration of the national government; the change in ownership, however, did not save it from destruction. In August 1907 the Kantō (関東) Region was hit by two typhoons, one right after the other, causing extensive damage. All rivers in Kanagawa (神奈川) and Tokyo areas were affected. The lower reaches of the Tama around Kawasaki overflowed its banks causing damage to houses, fields, embankments and roads. The nearby Tsurumi River (鶴見川) also flooded, turning the delta area into a great sea. Needless to say, the Rokugō Bridge crumbled. Worse flooding followed in 1910. For two weeks beginning on August 1 the Kantō Region experienced record rainfall, again aided by the arrival of two typhoons. All rivers in Kanagawa flooded causing massive damage through the prefecture. The Tama was some 7.6 meters above normal; newspapers reported: "the Keihin area [the area between Tokyo and Yokohama] has been turned into a sea (*Keihin wa umi to kasu* 京浜は海と化す)." Whole villages were washed away; fields and orchards were destroyed, embankments, bridges (including the Rokugō) and road damaged beyond repair (Figure 8.3).

The Great Flood of 1910 killed 1,231 persons nationwide and resulted in the loss of some 518,000 houses. Reports estimated the cost of the flood between four and five percent of the national product for that year.[14]

The unprecedented scale of the flooding spurred the central government to more aggressive attempts to control rivers such as the Tama. The River Law of 1896 focused on flood control and several large-scale projects were undertaken, including the building of select floodplains and embankments. Immediately after the 1910 flood, however, the central government created an Emergency Flood Control Committee to draw up strategic plans. Regarding the Tama, new laws were passed and monies from the national budget (7.4 million yen over ten years) allocated in 1911 to "re-form" the river. Construction work began in 1916. The lower reaches of the Tama River were straightened, banks reinforced with concrete, inlets and outlets of water into the main stream were cemented and placed under human control.

Figure 8.3 Postcard showing the destruction of the Rokugō Bridge after the Great Flood of 1910 (source: Author's collection. Courtesy of the Civil Engineering Library, Tokyo).

Bridges and roads

In 1918, plans were under way to build new national highways in Tokyo and Osaka. The establishment of the Road Improvement Association in March 1919 brought businessmen and politicians together in a "good roads movement" similar to the group that helped bring about the Federal Aid Road Act of 1916 in the United States. Headed by such eminent figures as Home Minister Tokunami Takejirō (床次竹二郎, 1866–1935), Matsukata Masayoshi (松方正義, 1835–1924; at that time Lord Keeper of the Privy Seal), and entrepreneur Shibuzawa Eiichi (渋沢栄一, 1840–1931), the association lobbied on behalf of a new road law to replace the antiquated regulations of 1877.

Road builder Samuel Hill (1857–1931), visiting Japan in January 1919, added his weight to the movement. In an essay published in March 1919, he professed himself shocked to learn that "There is not one mile of hard-surfaced road in the Empire." He went on to say that the poor condition of Japan's roads gave foreign visitors the impression of disorderliness and sordidness.

> The roads are dirt, and the frequent rains make them continually muddy. So that this mud may not become too deep small stones are strewn over the surface, which make walking especially disagreeable. There are no sidewalks anywhere. It is impossible to motor any distance with comfort, for there is always the possibility that one's trip may not be continued by reason

> of the lack of bridges over the various streams and creeks. What bridges there are too narrow for the gauge of a modern motor car. There are no graded roads.

He found it baffling that Japan, which has shown itself among the foremost in everything that pertains to progress, should nonetheless be "the most backward in highway building." A good roads movement, he maintained, would contribute greatly to Japan's prosperity:

> If people could only see what an important part road-building would play in its life, what it would mean in health, comfort and riches to its people, and could only understand that it is the most important problem which it has to face to-day, Japan would build roads. Japan before has experienced a period of prosperity due to its improved transportation by water and by railways, and just as those two things ministered to her economic advance so will highway building contribute to it, although to a far greater extent.[15]

As if in response to this message, on April 10, 1919, the national government enacted a much-needed new Road Law.[16] Together with a revised and toughened River Law, the Road Law (which went into effect in 1920) initiated a new era of large-scale civil engineering. The Home Ministry, envisioning Japan in the year 1950, issued a 30-year plan involving the paving of prefectural and national roads, the construction of new highways designed specifically for automobile traffic, the construction of permanent bridges, river embankments, dykes and dams. It floated a loan to raise 282.8 million yen to help pay for these massive public works projects. It is at this point that rivers, roads and bridges combined in the making of modern Japan.

The addition of motorized transport on Japanese roads added pressure to the good road movement. By the early twentieth century, Japan's streets, roads and bridges were dominated by a chaotic mix of pedestrians, bicycles, ox carts, horse-drawn carts and coaches, rickshaws and pushcarts, and a few, but growing number of passenger cars and trucks.[17] In 1920 there were only 7,912 cars and trucks (of which 2,695 were in Tokyo) in the entire country; while in the United States the number was 9.2 million.[18]

While numbers were small, the environmental and economic consequences of this emerging car culture were great, serving to initiate an "age of reconstruction" in the years before the 1923 Great Kanto Earthquake. Gravel, asphalt and concrete were in sudden demand for a surge in civil engineering projects, including highways and bridges. The example of Kanagawa Prefecture is instructive; between 1921 and 1926, cities, towns and villages built some 343,800 square meters of new road, improved some 22,843,000 square meters of existing road, and built or re-built 2,439 bridges.[19] Although not uniform, a similar explosion of road and bridge building took place in other prefectures. By the end of fiscal year 1925, some 8,228 kilometers of national roads, 93,094 kilometers of prefectural roads, 17,648 kilometers of municipal roads, and 920,220 kilometers of town and village roads had been completed at a cost of 126 million yen.[20]

The construction of the Keihin Highway (National Highway 1; now National Highway 15) between Tokyo and Yokohama featured the Rokugō Bridge across the Tama River as its centerpiece and provided a model for a future national road system.[21]

The Tokyo section of Japan's first hard-surface highway began at Yatsuyama (八ツ山) Bridge near Shinagawa (品川), where the old Tokaidō post road and the new Tokaidō rail line intersected, and capitalized on existing path in two different ways. An entirely new highway was built parallel to the post road as far as Suzugamori (鈴ヶ森), the location of the execution grounds that formerly greeted people as they entered Edo; from there, construction simply widened the Tokaidō until just before the Rokugō Crossing, where a new road led to the entrance of a new Rokugō Bridge. The Kanagawa portion of the highway beginning at the Kawasaki of the Tama was also new construction; rice fields were filled in and the ground leveled and compacted to create a road parallel to the Tokaidō rail line. At Namamugi (生麦) the highway rejoined the old Tokaidō post road that was widened, graded, and paved as far as Koyasu (Figure 8.4).

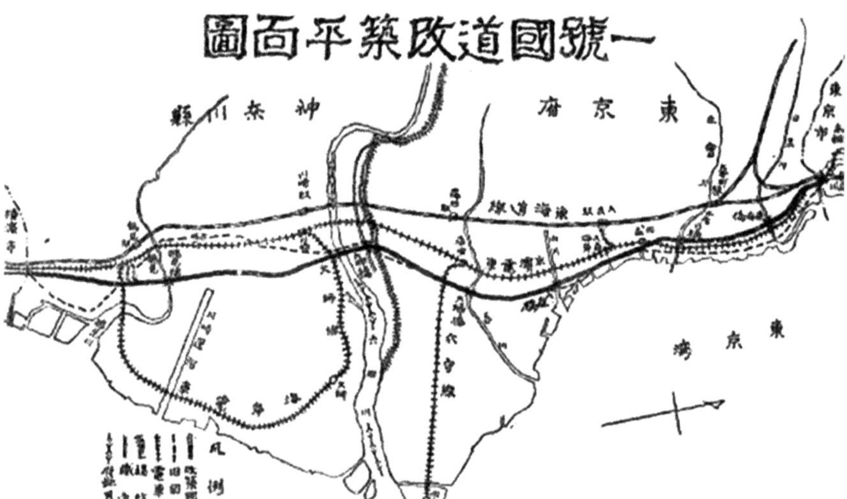

Figure 8.4 The Keihen Highway from Tokyo (Shinagawa) on the right to Yokohama on the left (source: *Dōro no kairyō*, 9.1 (January 1927). Courtesy of the Civil Engineering Library, Tokyo).

Note

The Tama River empties into Tokyo Bay at the center. The black and white line at the top indicates the path of the Tokaidō railroad; just beneath it are the tracks of the Keihin electric tram. The thick dark line is the path of the new Keihin Highway (National Highway 1); the route of the old Tokaidō as it diverges from the new route is shown by a broken line. The Rokugō Bridge is in the very center where the Keihin Highway crosses the Tama River.

Equipment imported from the United States and England included a variety of 8, 10, 12 and 15 ton rollers, graders, rock crushers, asphalt plants, concrete plants, asphalt pavers, dump trucks, cranes, pumps and gasoline-powered electric generators[22] (Figure 8.5).

The asphalt itself came from the Akita oil fields and gravel from Tama River pits. Cement was available from the giant Asano Cement Company and other domestic producers nearby. The width of the two-lane road varied between 18 and 27 meters, allowing cars and larger trucks to pass safely; pull-over areas were also created to improve safety. Careful attention was paid to gradient and camber, allowing rainwater to run off into covered culverts. Trees and other vegetation were planted on both sides of the road in the interests of beautification (Figure 8.6).

The Rokugō Bridge (Figure 8.7) was literally central to the project, at midpoint between the Tokyo and Kanagawa sections of the 27 kilometer highway. Design of the new bridge was entrusted to Masuda Jun (増田淳, 1883–1947), a civil engineer who became Japan's master bridge builder in the 1920s and 1930s. Born in 1883, he graduated from the Faculty of Engineering of Tokyo Imperial University in 1907.[23] University connections with John Alexander Low Waddell (1854–1938), an American bridge architect with teaching experience at the University of Tokyo in the 1880s and later author of major textbooks on bridge engineering, allowed him to take up a position in Hendrick and Conchrame (Consulting Engineers) in Kansas City, Missouri.[24] Masuda worked under Ira G. Hendrick for 15 years before returning to Japan in 1922 and setting up his own firm in the Gotanda section of Tokyo.[25]

The Rokugō Bridge was one of Masuda's first Japanese bridges; he drew up a meticulous set of plans in December 1922.[26] The bridge design called for a 135.3 meter tied arch span bridge (each span 67.655 meter) over the actual river channel on the Kanagawa side and a 310.9 meter girder bridge supported by 17 piers over the floodplain on the Tokyo side. The roadbed was 16.4 meters wide and consisted of reinforced concrete paved with asphalt, thus insuring ample space for motorized traffic, including large trucks.[27] Work was interrupted by the Great Kantō Earthquake in 1923 and flooding from typhoons that hit the Kantō area weeks after the earthquake, but construction proceeded smoothly thereafter, reaching completion in June 1925. The total cost was reported at 3.5 million yen. However, at this point it was a bridge going nowhere; the Tokyo section of the Keihin Highway was still under construction.

Nonetheless, great celebrations including religions ceremonies, speeches, and fireworks were held on the bridge to open both it and the Kanagawa section of the new highway to traffic on August 3, 1925.[28] Tokyo Governor Usami Katsuo (宇佐見勝, 1869–1942) praised the new Rokugō Bridge: "Now it has been given the status it deserves as the gateway to the metropolis and the most important bridge in the empire." He also praised the permanence of the construction of the new bridge "will ease the flow of people and goods between Tokyo and Yokohama, thereby to the good fortune not only of these two cities, but to the nation as a whole." Kanagawa Governor Seino Chōtarō (清野長太郎, 1869–1926)

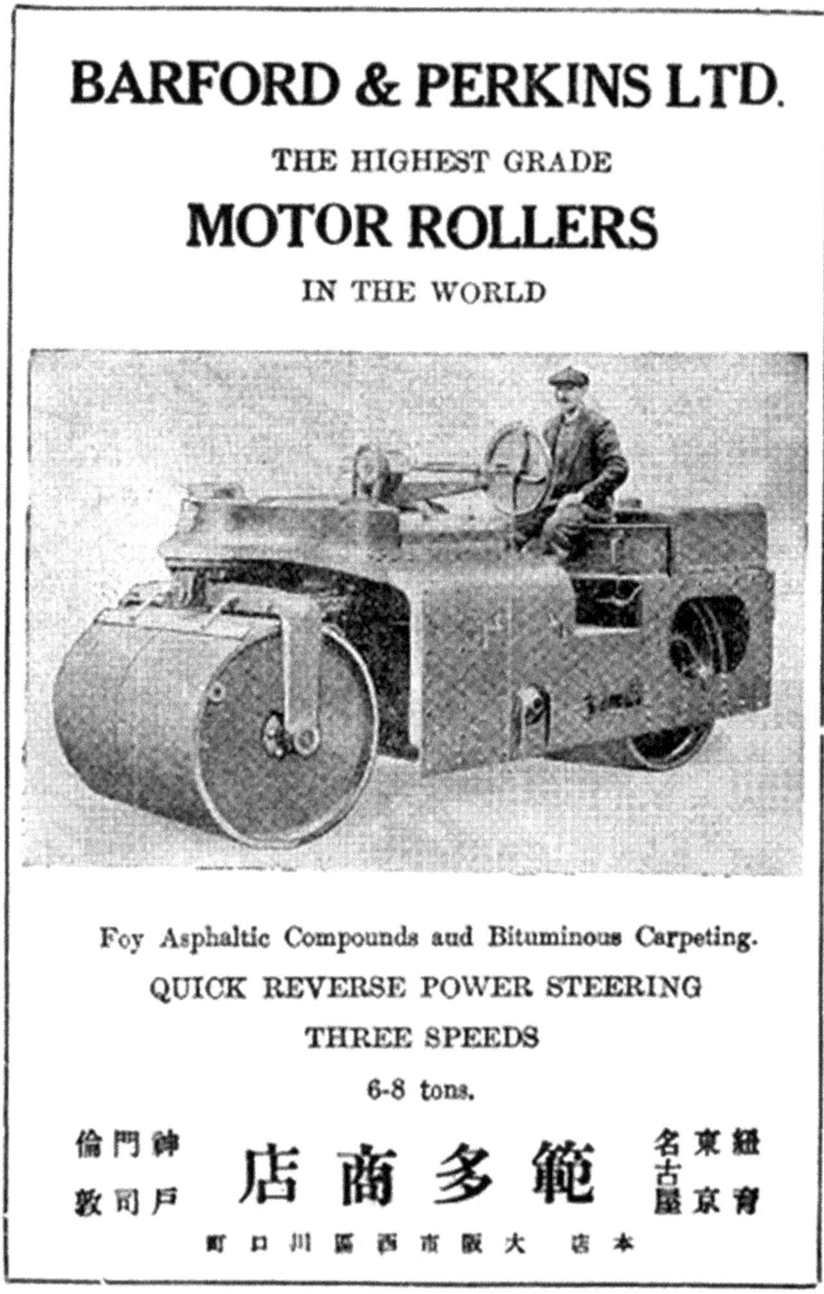

Figure 8.5 Advertisement for imported road construction equipment, front matter (source: *Road Improvement* [*Dōro no kairyo* 道路の改良], 9.1 (January, 1927). Courtesy of the Civil Engineering Library, Tokyo).

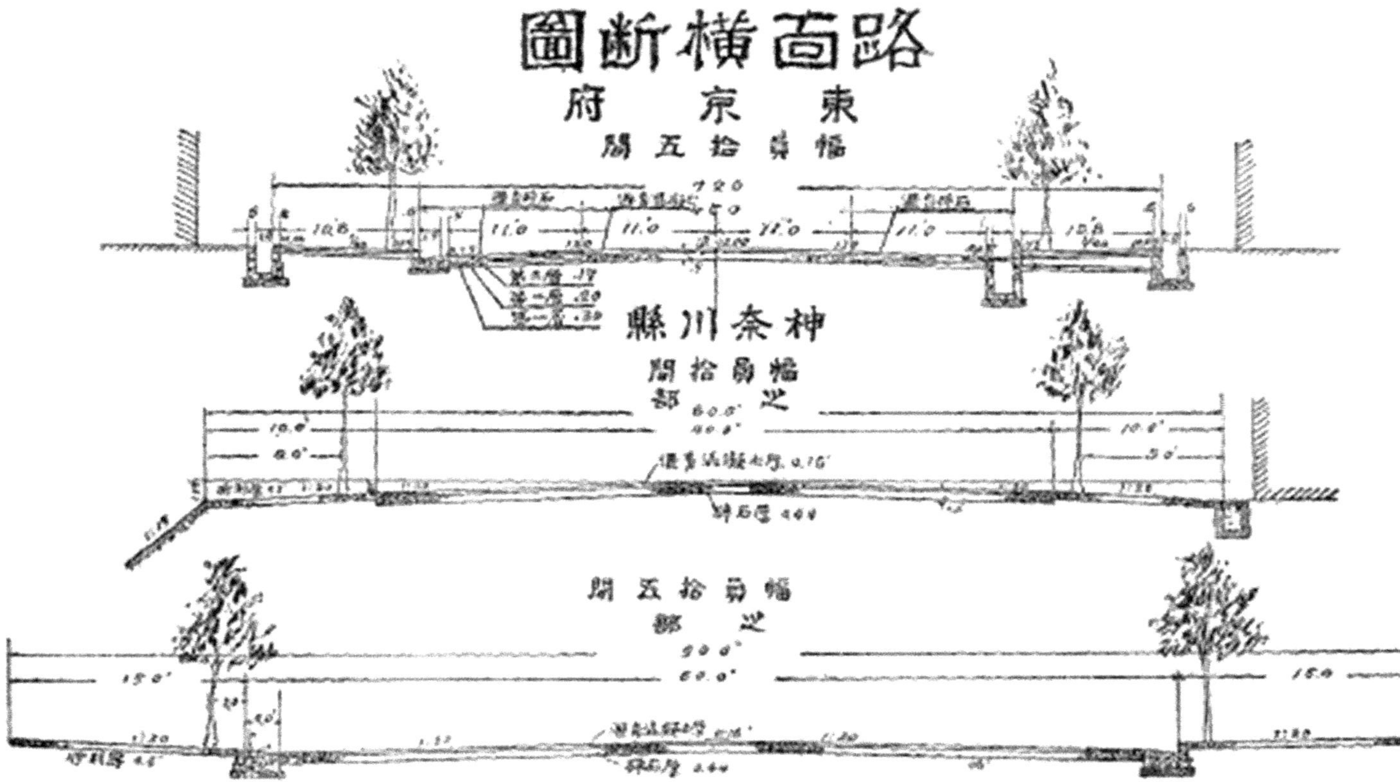

Figure 8.6 Cross-section of the new paved road; note calculations for grade and camber and trees on both sides (source: *Road Improvement* [*Dōro no kairyō* 道路の改良], 9.1 (January 1927)).

Figure 8.7 View of completed Rokugō Bridge (source: *Doboku kensetsu kōji gahō* [土木建設工事画報], 2.1 (January 1926). Courtesy of the Civil Engineering Library, Tokyo).

attributed even a greater future for the bridge. He began by recognizing the long history to the Rokugō Crossing, but noted that the ferry and the bridge were no longer suitable to the transportation needs of the new age. The construction of the new permanent bridge had been an urgent necessity. "Thanks to the beneficence of the government, we are blessed with a new bridge and a new highway without peer in Asia." He concluded by topping the claim made by the Tokyo governor: "This bridge will not only contribute to the convenience of the people of Kanagawa, but to the prosperity of the entire country—and indeed to the advancement of world culture." Messages from other dignitaries, including Prime Minister Katō Kōmei (加藤高明, 1860–1926), Home Minister Wakatsuki Reijirō (若槻禮次郎, 1866–1949) and Finance Minister Hamaguchi Osachi (濱口雄幸, 1870–1931), repeated this message: the bridge and the new national highway were about to completely transform (*isshin* 一新) Japan's transport and commerce network, resulting in a new age of prosperity and progress.

Finally, the Tokyo section of the Highway was completed in 1927, after many delays and revisions to the original plan. Survey work for the highway and bridge had begun in 1917 and orders for large equipment were made. The original goal was to have the highway completed in 1922, but financial, legal and technical problems, to say nothing of the terrible destruction caused by the Great Kantō Earthquake and fires in September 1923, delayed the project considerably. The Tokyo section was open to traffic one year after the Kanagawa section, on

November 28, 1926. Compromises were made along the way, tram lines interrupted the flow of traffic at several points, telegraph poles marred the beauty of the road and driver safety, and major crossroads remained unpaved; nonetheless, the great Tokaidō, a narrow, pot-hole ridden, often muddy road with no dependable bridge across the Tama, was finally replaced by a broad concrete highway, allowing cars and trucks, people and goods, to speed between the two great cities of Tokyo and Yokohama, adding not only to the prosperity of these two cities, but of the entire country. Such was the rhetoric of the congratulatory speeches given at the second opening ceremony on November 28, 1926.[29] After the speeches, a round of toasts and banzais were held in the expectation that the reborn Tokaidō would advance commerce, industry, culture and, befitting the new car culture, happiness.[30]

Conclusion—a concrete legacy

The 1920s continued a golden age of modern ferro-concrete bridge building in Japan. Masuda Jun was responsible for over 50 such bridges during the ten years between 1922 and 1932.[31] The Rokugō Bridge was built to withstand the ravages of the Tama River. It remained in service until 1984 when it was replaced by a new bridge so seamless with other sections of that highway that drivers were hardly aware of crossing over a river that had once been deemed unbridgeable.

The magic of ferro-concrete construction played a role in taming the Tama, but ironically the new strength of the bridge was complemented by changes in the river that significantly diminished its destructive force. Beginning in the early twentieth century, the proven superiority of reinforced concrete roads, bridges, docks and buildings meant that new construction projects involved vast quantities of concrete, which in turn required vast quantities of gravel, much of it dug out of the Tama River.[32] This was especially true after the Great Kantō Earthquake of September 1, 1923.

After the Great Kantō Earthquake, demand for gravel continued to rise. Between 1912 and 1915, an estimated 86,250 tons of gravel were taken from the Tama River. In the period 1916–1919, this figure grew to 115,000 tons. In 1922, 1,150,000 tons of gravel was extracted from the Tama, a ten-fold increase over 1919. After the Great Kanto Earthquake, and the massive reconstruction projects to rebuild Tokyo and Yokohama, demand for gravel continued to rise; in 1925 the figure was 1,400,000 tons and in 1925, 1,450,000 tons.[33]

By the 1920s, the gravel industry was no longer a cause of flooding; quite the reverse, the removal of rocks had significantly lowered the water level, in places exposing the foundations of bridge supports. In the absence of stones and gravel, the river emptied into Tokyo Bay without impediments to slow its flow.[34] The "gentle, deep" river was no more. Shallow waters brought river traffic to an end. Ironically, by 1925, when the Rokugō Bridge was completed, the Tama and many other rivers, robbed of their rocks and cemented in place, were no longer a menace to their bridges.

Moreover, despite the many accolades, the Rokugo Bridge and the Keihin National Highway failed to immediately open up a new age of enhanced land traffic convenience. A satirical cartoon published by *Road Improvement* [道路の改良] described key events in Japan's good road movement, month by month for the year 1926: in October Home Minster Hamaguchi cut the budget for road building; in November the Keihin Highway opened, but the new road came to a sudden end at Yatsuyama Bridge near Shinagawa; in November the Hanshin Highway (National Highway 2) was opened, but only with one lane[35] (Figure 8.8).

A cartoon for January 1927 reported that the government had passed legislation to build highways for automobile use, but the illustration showed how piecemeal construction would be.[36] An earlier (1925) cartoon lampooned the promise of the new Rokugō Bridge. Motorists seeking speed and convenience were surely disappointed when they realized that they had to share the road with horse carts, rickshaws, bicycles and pedestrians[37] (Figure 8.9).

Automobile ownership grew spectacularly in the interwar years, from just under 8,000 in 1920 to 66,306 automobiles in 1927, reaching a peak in 1938 at 221,162 before declining in the war years. But the good road movement failed to take off; only the Keihin National Highway and its Kansai equivalent, the 26.8 kilometer Hanshin National Road (National Highway 2) completed in 1928, linking Osaka with Nishinomiya, were built specifically with automobile traffic in mind.[38] By 1940, only about 1.8 percent of all roads were paved; the national highway system was somewhat better at 18.3 percent but that amounted to only 1,600 kilometers of pavement.[39]

What happened? First, Japanese people "on the move" had access to a variety of inexpensive all-terrain transport, especially bicycles and rickshaws. Second, the cost of good roads and bridges and other public works limited highway expansion. The reconstruction of Tokyo and Yokohama following the 1923 earthquake depressed the economy from the mid-1920s, well before the onset of the world depression in 1929. Finally, different priorities were in work. In the depression years of the 1930s, the government relied on military spending (as opposed to public works such as roads and dams) as a means to stimulate the economy. Plans drawn up in the late 1930s to build a Japanese equivalent of Germany's autobahn failed to materialize. There were even more grandiose plans for a Great East Asian Highway, but aside from some work done around Nagoya and Kobe, the war in China and later in the Pacific effectively shelved plans for highway construction.[40]

Nonetheless, the years between 1918 and 1932, the golden years of ferro-concrete construction and the take-off of motorized transport, did leave a powerful, although ambiguous, legacy to postwar Japan. First was the promise of unrestricted mobility. Beginning in the late 1950s, construction began on a vast system of national expressways, national toll roads and metropolitan expressways. The number of private automobiles expanded rapidly from 139,000 in 1955 to 1.7 million ten years later. By 1975, the number had shot up to 16 million.[41]

Second, the so-called "economic miracle" of Japan's postwar recovery in the 1950s and 1960s was the cause of a second construction boom and renewed

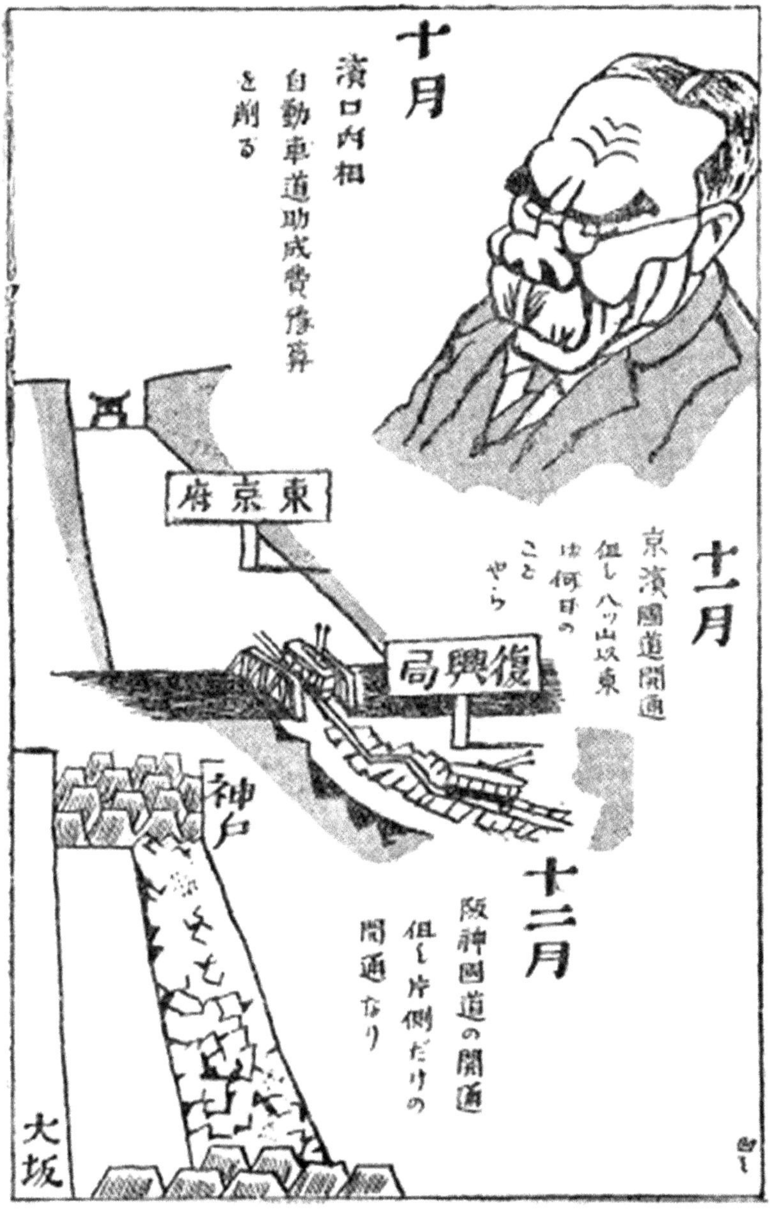

Figure 8.8 A cartoon history for 1926. The captions read, top to bottom: October: The Hamaguchi Cabinet cuts the highway budget. November: The Tokyo–Yokohama Highway opened, but when will there be a road east of Yatsuyama? December: The Osaka–Kobe Highway opened, but with only one lane (source: *Road Improvement* [*Dōro no kairyo* 道路の改良], 9.1 (January, 1927). Courtesy of the Civil Engineering Library, Tokyo).

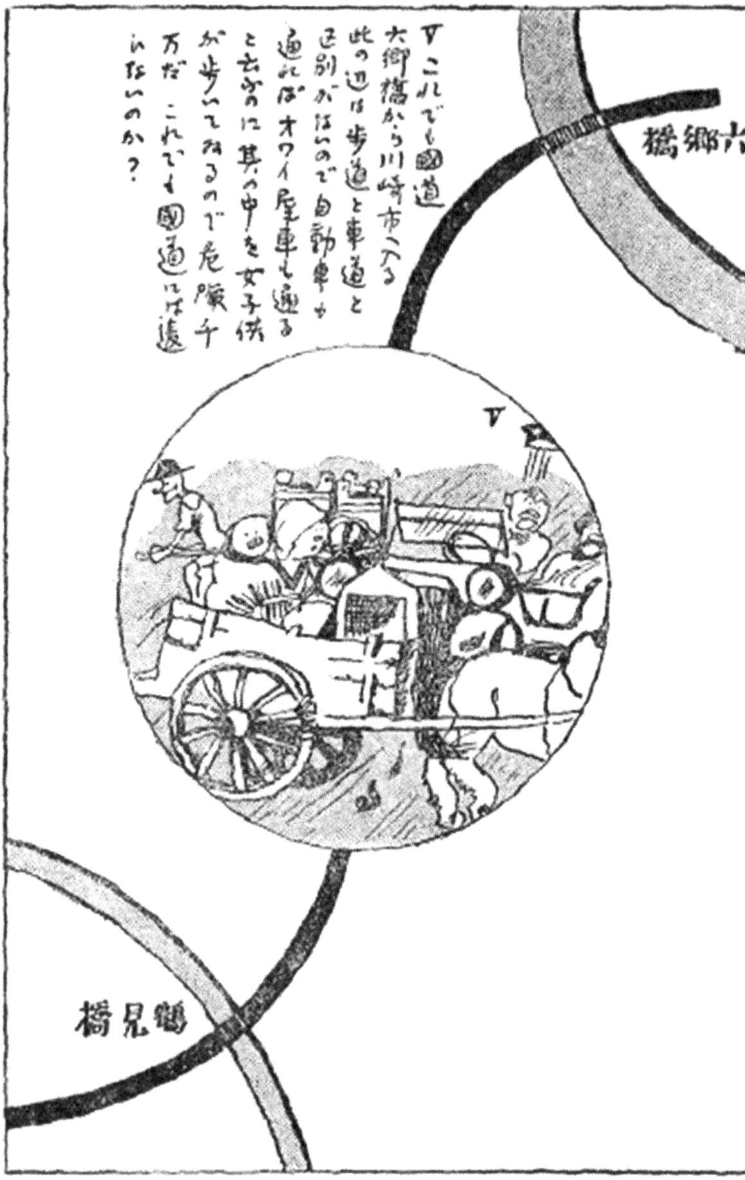

Figure 8.9 Travel sketches between Shinagawa and Kanagawa: The opening of the Rokugō Bridge. The caption reads: "Upon crossing the Rokugō Bridge into Kawasaki there is no distinction between the pedestrian path and the road for cars. Automobiles must compete with night soil carts and with mothers on foot with babies on their back. This is dangerous. Does this really qualify as a National Highway?" (source: *Road Improvement* [*Dōro no kairyo* 道路の改良] 7.11 (November, 1925). Courtesy of the Civil Engineering Library, Tokyo).

demand for gravel from the Tama and other rivers. In addition to new ferro-concrete buildings, bridges and road, concrete was used for river and coastline repair, tetrapods, embankments and dams. These massive construction projects all required gravel and other mineral resources including sand and limestone. Tama River gravel pits were reopened in 1946. Throughout the 1950s, some 600,000 tons of gravel were dug out of the Tama River yearly. By the early 1960s, however, damage to the river caused the government to impose restrictions on excavation, and by 1968, river gravel collection had come nearly to a halt, replaced by mountain gravel, crushed rock and imported materials.[42] The cement industry also prospered in the immediate postwar years. The prewar production of cement peaked in 1939 at 6,500,000 tons.[43] During the war years cement production declined dramatically, but the postwar construction boom revived the industry. The prewar peak was achieved in 1952; by 1956 Japan was producing 13,737,594 tons of cement. Ten years later, in 1966, this figure had nearly tripled at 38,567,179 tons.[44] For better or worse, the "construction state" is another legacy of the years that saw the birth of the Rokugō Bridge and the Keihin National Highway.

The construction industry has continued to exert a strong force on Japanese political and economic decision making, especially under the years when the Liberal Democratic Party (LDP) was in power—which is nearly the entire postwar period, making Japan into one of the most concrete covered nations in the world. When the opposition Democratic Party of Japan (DJP) came briefly to power in 2009, the new Prime Minister Hatoyama Yukio (鳩山由紀夫, 1947–) attempted to shift spending priorities "from concrete to people," placing emphasis on social welfare. His administration vowed to halt construction of the Yamba Dam in Gunma Prefecture, even though 70 percent of the 460 billion yen project had already been spent.[45] The LDP regained control of government in 2012 and reversed this decision; the massive concrete dam is currently scheduled for completion in 2019. Coupled with the reconstruction needs following the Great Northern Japan Earthquake and Tsunami of March 11, 2011, cement production has surged. According to a May 13, 2013 article in the *Japan Times*, there is a soaring demand for concrete for roads, sea-walls and buildings in the disaster-affected areas. "Ships are bringing gravel from Aomori and Hokkaido. Concrete factories have taken on staff and bought more trucks. Production is going at full tilt."[46]

The LDP has announced plans to submit a bill to the Diet to boost spending on infrastructure nationwide. It calls for ¥200 trillion investment over ten years in tsunami-proof cities, new roads and better sewers, among other things, to protect against natural and manmade disasters. On the one hand this smacks of a tradition of pork barrel spending on concrete that goes back to the 1920s; on the other hand we see here colossal pollution and disfigurement of the natural environment, the inexorable results of what Alex Kerr terms "a systematic addiction to construction."[47] But more fundamentally, concrete itself is an instrument of modernity, replete with contradictions—and this too has its origins in the "golden age" of the 1920s. As Adrian Forty has noted,

> What makes concrete "modern" was that while it has been liberating, it has at the same time been destructive. Its structural possibilities allow us to redirect our destiny, to build bridges and superhighways where none could be built before, to hold back the sea, to join continents, to build cities at densities which would previously have resulted only in unbearable squalor, to reverse or to overcome the forces of nature so that mankind can live in greater comfort and security. In all these senses, concrete has been emancipatory.

But Forty also notes that

> To "concrete over" is to end nature's dominion over the surface of the earth. As well as bringing us closer together, it cut us off from one another.... And these changes are permanent. With concrete, there is no going back. Its indestructibility is both one of its most valued, and at the same time most reviled features.[48]

To be sure, concrete and the bridges and roads constructed from it, comes at a cost to the environment; but there is no turning back. At best, learning from history, we can soften the blow.

Notes

1 The major source of information on the Tama River is the multi-volume *History of the Tama River* [多摩川誌], published by the Foundation of River and Watershed Environment Management 河川環境管理財団, in 1986 and revised in 1994. Volume 1 is primarily devoted to the history of the Tama River. A shorter account may be found in Miwa Shuzō 三輪修三, *Tamagawa* [多摩川], (Yokohama: Yurindō, 1988).

2 Engelbert Kampfer, *Kaempfer's Japan: Tokugawa Culture Observed*, edited and translated by Beatrice Bodart-Bailey (Honolulu: University of Hawaii Press, 1999), p. 348.

3 Miwa, *Tamagawa*, pp. 66–67. See also Constantine Vaporis, *Breaking Barriers: Travel and the State in Early Modern Japan* (Cambridge, MA: Harvard University Press, 1994), p. 52.

4 On the travels of the elephant in 1629, see Oba Osamu, Part Eight: "The Travels of Elephants," *Sino-Japanese Relations in the Edo Period*, pp. 50–68, available online: www.chinajapan.org/articles/11.2/11.2oba50–68.pdf, accessed on January 20, 2015.

5 *History of the Tama River*, Vol. 1, pp. 251–252, 266–268.

6 *History of the Tama River*, Vol. 1, pp. 279–280.

7 C. J. Van Doorn from Holland remained in Japan from 1872 to 1880; he worked on taming the Yoko River and Tone River and was responsible for introducing Western technology relating to water management. Other foreign engineers included I. H. Lindow (in Japan 1872–1875) and Johanes D'ijke (in Japan 1873–1901).

8 Biographical information on Edmund Morel may be found in Ian Nish, ed., *Britain and Japan: Biographical Portraits* (London: Routledge, 1997), Vol. 2, "Edmund Morel, Pioneer Railroad Builder in Meiji Japan." For general information on the Rokugō River Railroad Bridge, see *The History of Kawasaki City, Narrative* [川崎市史·通史編], Vol. 2 (Kawasaki: Kawasaki-shi, 1994), pp. 126–130.

9 Boyle published a detailed account of the construction of the new Rokugō Railroad Bridge: "The Rokugo River Bridge and Foundations on the Tokio–Yokohama Railway, Japan," *Minutes of Proceedings of the Institution of Civil Engineers*, (1882),

pp. 216–228. It included an account of the Rokugō River Flood of September 16, 1878 on p. 228. Available online: http://book.google.cp.jp/books?id =MhIFAAAAQAAJ&pg=PA453&dq=boyle+rokugo+river+bridge&hl=ja&sa=X&ei=aDv3UeDEI46akgWMkYCYAQ&ved=0CDcQ6AEwAQ#v=onepage&q=boyle%20rokugo%20river%20bri&f=false, accessed on August 23, 2013. Finally, the 1885 *American Councilar Reports* contains an interesting description of the Rokugō River and the railway bridge:

> The Rokugo River, about half-way between the termini, is at times a deep and turbulent stream, overflowing its banks with the rush of waters from the mountains, and the railway therefore required at this point a bridge of the very best construction. This was finished a few years since, in iron, by English engineers, and is an admirable piece of work, the trains crossing it at full speed. The cost of this bridge is stated to me by the bureau as 330,470.75 yen, but common rumor places it at about three times that amount.

"Railways in Japan," Report by Consul-General Van Buren, available online: http://books.Google.co.jp/books?id=to08AQAAIAAJ&pg=PA688&dq=rokugo+river+bridge&hl=ja&sa= TEj3UdzWNcHpkAXRi4DABw&ved=0CGIQ6AEwBzgo#v=onepage&q=rokugo%20river20bridge&f=true, accessed on December 31, 2013.

10 "The New Railroad Bridge at Kawasaki," *Japan Weekly Mail*, October 6, 1877. This article also included technical details on the new bridge:

> The superstructure of the bridge proper consists of six spans, each of two wrought iron girders, of the kind called "Warren's girders," 100 feet in length and 12 feet in height, between which are laid smaller cross girders 27 feet in length, upon which again beams of timber are laid at right angles and parallel to the main girders, for carrying a double line of rails of 3 feet 6 inch gauge; leaving an intermediate space between the up and down lines of 6 feet, the whole width of roadway between the main girders being 22 feet. The abutment upon the Yokohama side of the river, upon which the ends of the first pair of girders are laid, consists of four brick wells built together in a group connected by arches of brickwork and capped with massive blocks of granite, laid so as to form a solid pier of great strength. Each of the four brick wells is of 12 feet in diameter, the circular brick work between 2 feet in thickness, consisting of bricks specially made for the purpose and laid in Portland and Selenitic cement, with iron tie rods running through from bottom to top in order to increase the strength of the whole mass, and secure a firm bond.

http://books.google.co.jp/books?id=MhIFAAAAQAAJ&pg=PA453&dq=boyle+rokugo+river+bridge&hl=ja&sa=X&ei=aDv3UeDEI46akgWMkYCYAQ&ved=0CDcQ6AEwAQ#v=onepage&q=boyle%20rokugo%20river%20bri&f=false, accessed on August 23, 2013.

11 On Suzuki, see E. G. Seidensticker, *High City Low City*, (Rutland, VT: Tuttle Publishers, 1984), p. 58. On Sanai and his attempt to bridge the Tama, see *The History of Kawasaki City, Narrative*, Vol. 2, pp. 144–148, and *The History of Kanagawa Prefecture, Narrative* [神奈川県史・通史編], Vol. 6, (Yokohama: Kanagawa-ken, 1982), pp. 145–146.

12 *The History of Kanagawa Prefecture, Narrative*, Vol. 6, pp. 146–147.

13 For a detailed and illustrated history of tram service over the Tama River, see *The History of the Keikyū Tram Bridge at Rokugō* [京急六郷鉄道橋の歴史], available online: www.tamagawa-kisui.jp/ref/ref6/ref-6.html, accessed on August 23, 2013.

14 Seidensticker, *High City Low City*, p. 58.

15 M. A. Salomon, "A Good Roads Movement in Japan: One Way in which American Efficiency Is Helping the Flowery Kingdom," *The Outlook*, March 26, 1919, pp. 12–13. Available online: www.unz.org/Pub/Outlook-1919mar26–00515, accessed

on January 20, 2015. A transcript of Hill's speech (in Japanese translation) to the Japan Good Roads Association on May 5, 1920 appeared in the inaugural issue of *Road Improvement* [道路の改良], Vol. 1, November 1920 under the title "On Constructing Better Roads (善良なる道路の築造)." Available online: http://library.jsce.or.jp/Image_DB/mag/dokai/pdf/01–01–0033.pdf, accessed on January 20, 2015.

16 See Yamamoto Hirofumi, Chapter 4, "Developing an Independent Transportation Technology (1910–1921): Roads," in Yamamoto Hirofumi, ed., *Technological Innovation and the Development of Transportation in Japan* (Tokyo: United Nations University Press, 1993), pp. 96–101, on the importance of the Road Law of 1919, available online: http://d-arch.ide.go.jp/je_archive/pdf/book/jes6_a.pdf, accessed on January 20, 2015; see also *An Outline History of Road and Traffic Policy* [道路交通政策史概要], a website compiled by the Association for the Study of Road and Traffic Problems 道路交通問題研究会. This work provides basic information on traffic legislation, including the Road Law of 1919; see especially Chapter 2, available online: www.taikasha.com/doko/gaikancon1.htm, accessed on January 20, 2015.

17 See Yamamoto, ed., *Technological Innovation*, "Table 5. Total Vehicles 1910–1939," p. 97. The chart noted that between 1910 and 1930 the number of horse-drawn passenger carriages declined from 8,565 to 2,175, however the number of freight horse-drawn carriages increased from 158,590 to 308,914. Ox carts similarly rose in number from 35,448 to 98,690. The number of rickshaws declined from 149,567 to 42,635 whereas the number of bicycles rose rapidly from 239,474 to 5,779,297. The first reliable automobile census shows 873 passenger automobiles in 1915, growing to 57,827 in 1930.

18 For Japan, see Yamamoto, ed., *Technological Innovation*, p. 97; for the United States, see statistics on "Motor Vehicle Factory Sales and Registrations, and Motor-Fuel Usage, 1900–1957," in United States Bureau of the Census (ed.), *Historical Statistics of the United States, Colonial Times to 1957*, U.S. Department of Commerce, 1960, p. 462. The total number of automobiles and trucks registered in 1920 was 9,229,141.

19 Yamamoto, ed., *Technological Innovation*, p. 99.

20 Publications Committee, World Engineering Congress, *Industrial Japan: A Collection of Papers by Specialists on Various Branches of Industry in Japan*, (Tokyo: International Publishing and Printing Co., 1929), p. 87.

21 Ibid., pp. 86–87.

22 For details, see Tanaka Yoshimi 田中好, "Improvements on the Keihin Highway (改良された京浜国道)," *Road Improvement* [道路の改良], 9.1 (January 1927), pp. 145–146 for equipment newly purchased for the Tokyo portion of the road and pp. 142–146 for the Kanagawa portion.

23 There is only scant biographical information on Masuda Jun. See Fukui Jirō 福井治郎 "Souvenir of Jiun Masuda, Great Bridge Designer (橋梁設計事実者増田淳の足跡)," *Papers on Civil Engineering* [土木研究論文集], Vol. 23 (2004), pp. 165–175, available online: https://www.jstage.jst.go.jp/article/journalhs2004/23/0/23_0_165/_pdf, accessed on January 20, 2015. A catalogue of the Masuda Jun papers, housed in the Tsukuba Research Institute of Civil Engineering, are available online: http://library.jsce.or.jp/jscelib/committee/2006/draw/masuda/masuda.htm, accessed on January 20, 2015.

24 See the Wikipedia biography of John Waddell: http://en.wikipedia.org/wiki/John_Alexander_Low_Waddell; the Japanese Wikipedia entry includes more information on his Japanese connections: http://ja.wikipedia.org/wiki/ジョン・ワデル, accessed on January 20, 2015.

25 A comprehensive list of Masuda's bridge works compiled by Yoshioka Atsushi of the Tsukuba Research Institute of Civil Engineering is available online: www.pwri.go.jp/jpn/seika/pdf/report-seika/2008–4–1–2.pdf, accessed on January 20, 2015. In the United States Masuda was responsible for the design and construction of some 30

bridges in 12 different states; after returning to Japan in the period between 1922 and 1931 he designed 55 bridges throughout Japan and in Korea and Taiwan. Detailed construction plans for many of his bridges have been preserved online at: http://library.jsce.or.jp/jscelib/committee/2006/draw/masuda/masuda.htm, accessed on January 20, 2015.

26 For the "Calculation Paper on the Design of Tied Span Arch Bridge in Rokugō Bridge," see: http://library.jsce.or.jp/jscelib/committee/2006/draw/masuda/20.htm, accessed on January 20, 2015.

27 For construction details and photo montage of the Rokugō Bridge, see Takada Kei 高田景, "The Rokugō Bridge, the Connecting Link in the New Tokyo Yokohama Road, Now Nearing Completion (京浜国道六郷橋と鶴見橋)," *An Illustrated Construction Review* [土木建築工事画報], 2.1 (January 1926), pp. 31–34. Available online: http://library.jsce.or.jp/Image_DB/mag/gaho/kenchikukouji/02–01/02–01–0229.pdf, accessed on January 20, 2015.

28 For a comprehensive report on the opening ceremonies of the Kanagawa section of the new national highway, including transcripts of the congratulatory speeches, see: "Ceremony to Commemorate the Completion of the Kanagawa Section of the Keihin Highway (京浜国道神奈川県管内改築工事竣工式)," *Road Improvement* [道路の改良] 7.8 (August 1925), pp. 113–118. Available online: http://library.jsce.or.jp/Image_DB/mag/dokai/pdf/07–08–0108.pdf, accessed on January 20, 2015.

29 For the report of the November 28, 1926 ceremonies: *Road Improvement* [道路の改良] 9.1 (January 1927), pp. 197–202. Available online: http://library.jsce.or.jp/Image_DB/mag/dokai/pdf/09–01–0197.pdf, accessed on January 20, 2015.

30 Ibid., p. 201.

31 Many of these bridges are still in use, including the Juso Bridge over the Yodo River, the Yoshino River Bridge in Tokushima, the Muko River Bridge in Hyogo, the Ise Ohashi Bridge over the Nagara River and the Shirahige Bridge over the Sumida River in downtown Tokyo. For a photo montage of Masuda's various bridges, see www.kyudou.org/ijinden/ch4_masuda_p.pdf, accessed on January 20, 2015.

32 For more details on the gravel industry and the Tama River, see M. William Steele, "The History of the Tama River: Social Reconstructions," in T. Tvedt and E. Jakobsson, eds., *A History of Water*, Vol. 1: Water Control and River Biographies, (London: I. B. Tauris, 2006), pp. 217–238.

33 *History of the Tama River*, p. 979.

34 Details on the environmental impact of gravel extraction can be found in *Ota Ward History, Documents on Folklore* [大田区史·資料編·民俗], Ōta-ku Shi Hensan Iinkai, ed., (Tokyo: Ōta-ku, 1983), pp. 68–71.

35 "A Cartoon History for 1926 (太正十五年漫画史)," *Road Improvement* [道路の改良] 9.1 (January, 1927). Available online: http://library.jsce.or.jp/Image_DB/mag/dokai/pdf/09–01–0190.pdf, accessed on January 20, 2015.

36 "A Cartoon History of Road Policy for 1927 (昭和二年路政漫画史)," *Road Improvement* [道路の改良] 10.1 (January, 1928). Available online: http://library.jsce.or.jp/Image_DB/mag/dokai/pdf/10–01–0188.pdf, accessed on January 20, 2015.

37 "A Cartoon Sketch of Road Life between Shinagawa and Kanagawa (Manga— Shinagawa Kanagawa kan suketchi dōchūki (漫画—品川神奈川間スケッチ道中記)," *Road Improvement* [道路の改良] 7.11 (November, 1925). Available online: http://library.jsce.or.jp/Image_DB/mag/dokai/pdf/07–11–0054.pdf, accessed on January 20, 2015.

38 The 210 meter long bridge across the Muko River at midpoint between Osaka and Nishinomiya was also designed by Masuda Jun. Photo montage is available online: http://hardcandy.exblog.jp/4488394/, accessed on January 20, 2015.

39 See the section on prewar road pavement in *An Outline History of Road and Traffic Policy* [*Dōro kōtsu seisaku-shi gaiyō* 道路交通政策史概要], available online: www.taikasha.com/doko/chapt11.htm, accessed on January 20, 2015.

40 For more details on the Great East Asian Highway, see Takebe Kenichi 高部健一, *A Talk about Road* [道の話し], (Tokyo: Gihōdō Shuppan, 2000), Vol. 1, pp. 135–140.
41 For a study of postwar mobility, comparing the growth of bicycle and automobile transport, see M. William Steele, "The Making of a Bicycle Nation: Japan," *Transfers*, 2.2 (Summer 2012). See especially pp. 76–79 for information on transport and roads in the 1960s. As of 2013, automobile ownership was just under 60 million units.
42 *History of the Tama River*, pp. 995–1002.
43 Doboku Gakkai, ed., *A History of Civil Engineering in Japan, 1912–1940* [日本土木史 1912–1940], (Tokyo: Doboku Gakkai, 1965), p. 1623.
44 Doboku Gakkai, ed., *A History of Civil Engineering in Japan, 1941–1965* [日本土木史, 1941–1965], (Tokyo: Doboku Gakkai, 1973), p. 152. For further information on the cement industry in Japan, see Nagata Shirō 永田四郎, *A History of the Cement Industry in Japan* [日本セメント産業史], (Kensetsu Bunkasha, 1957), and Yamada Junji 山田順治, *A Cultural History of Concrete* [コンクリートの文化史], (Tokyo: Bunichi Sōgō Shuppan, 1986).
45

> Boasting the motto to shift spending priorities "from concrete to people," the 2010 fiscal budget currently undergoing Diet deliberations looks to up welfare spending nearly 10 percent to ¥27.27 trillion, while slashing public works projects by more than 18 percent to ¥5.77 trillion, the lowest level in 32 years.

See "Half-year on, Hatoyama struggling," *Japan Times*, March 16, 2010, available online: www.japantimes.co.jp/news/2010/03/16/national/half-year-on-hatoyama-struggling/#.UgoPDRb8ETM, accessed on January 20, 2015.
46 "The Other Costs of Cement," *Japan Times*, May 19, 2013, available online: www.japantimes.co.jp/life/2013/05/19/environment/the-other-costs-of-concrete/#.Ugn4KR-b8ETM, accessed on January 20, 2015.
47 See Adam Kerr's "Conclusion," *Dogs and Demons: Tales from the Dark Side of Japan*, (New York: Hill and Wang, 2001), pp. 382–386.
48 Adrian Forty, "Concrete and Memory," in Mark Crinson, ed., *Urban Memory: History and Amnesia in the Modern City*, (London: Routledge, 2005), pp. 79–80.

Bibliography

Boyle, R. V. "The Rokugo River Bridge and Foundations on the Tokio-Yokohama Railway, Japan," *Minutes of Proceedings of the Institution of Civil Engineers*, 1882, pp. 216–228. Available online: http://books.google.co.jp/books?id=MhIFAAAAQAAJ&pg=PA453&dq=boyle+rokugo+river+bridge&hl=ja&sa=X&ei=aDv3UeDEI46akgWMkYCYAQ&ved=0CDcQ6AEwAQ#v=onepage&q=boyle%20rokugo%20river%20bri&f=false, accessed May 25, 2015.

Doboku Gakkai 土木学会 (ed.). *Nihon dobokushi* 日本土木史1912–1940 [*A History of Civil Engineering in Japan, 1912–1940*], Tokyo: Doboku Gakkai, 1965.

Doboku Gakkai 土木学会 (ed.). *Nihon dobokushi* 日本土木史1941–1965 [*A History of Civil Engineering in Japan, 1941–1965*], Tokyo: Doboku Gakkai, 1973.

Dōro kōtsū seisaku-shi gaiyō 道路交通政策史概要 [*An Outline History of Road and Traffic Policy*], a website compiled by the Dōro Kōtsū Mondai Kenkyū-kai 道路交通問題研究会 (Association for the Study of Road and Traffic Problems): www.taikasha.com/doko/gaikancon1.htm, accessed May 25, 2015.

Forty, Adrian. "Concrete and Memory," in Mark Crinson, ed., *Urban Memory: History and Amnesia in the Modern City*, London: Routledge, 2005, pp. 79–80.

Fukui Jirō 福井治郎. "Kyōryō sekkei jijitsu-sha Masuda Jun no ashiato 橋梁設計事実者増田淳の足跡 (Souvenir of Jiun Masuda, Great Bridge Designer)," *Doboku kenkyū ronbunshū* 土木研究論文集 [*Papers on Civil Engineering*], Vol. 23 (2004): 165–175. Available online: http://library.jsce.or.jp/jsce/open/00044/2004/23-0165.pdf, accessed May 25, 2015.

"Half-year on, Hatoyama Struggling," *Japan Times*, March 16, 2010. Available online: www.japantimes.co.jp/news/2010/03/16/national/half-year-on-hatoyama-struggling/#. UgoPDRb8ETM, accessed May 25, 2015.

Hill, Samuel, translation of his appeal to improve roads in Japan: "*Zenryō naru dorō no chikuzō* 善良なる道路の築造 (On Constructing Better Roads)," *Dōro no kairyō* 道路の改良 [*Road Improvement*], Vol. 1, November 1920. Available online: http://library.jsce.or.jp/Image_DB/mag/dokai/pdf/01-01-0033.pdf, accessed May 25, 2015.

Kampfer, Engelbert. *Kaempfer's Japan: Tokugawa Culture Observed*, edited and translated by Beatrice Bodart-Bailey, Honolulu: University of Hawaii Press, 1999.

Kanagawa-ken shi, tsūshi-hen 神奈川県史・通史編 [*History of Kanagawa Prefecture, Narrative*], Yokohama: Kanagawa-ken, 1982.

Kawasaki shishi, tsūshihen 川崎市史・通史編 [*Kawasaki City History, Narrative*], Kawasaki: Kawasaki-shi, 1994.

"Keihin kokudō Kanagawa-ken kannai kaichiku kōji shunkō shiki 京浜国道神奈川県管内工事竣工式 (Ceremony to Commemorate the Completion of the Kanagawa Section of the Keihin Highway)," in *Dōro kairyō*, 7.8 (August 1925): 113–118. Available online: http://library.jsce.or.jp/Image_DB/mag/dokai/pdf/07-08-0108.pdf, accessed May 25, 2015.

Keikyū Rokugō tetsudō hashi no rekishi 京急六郷鉄道橋の歴史 (A History of the Keikyū, Tram Bridge at Rokugō). Available online: www.tamagawa-kisui.jp/ref/ref6/ref-6.html, accessed May 25, 2015.

Kerr, Adam. "Conclusion," in *Dogs and Demons: Tales from the Dark Side of Japan*, New York: Hill and Wang, 2001, pp. 382–386.

"Manga—Shinagawa Kanagawa kan suketchi dōchūki 漫画品川神奈川間スケッチ道中記 (A Cartoon Sketch of Road Life between Shinagawa and Kanagawa)," *Dōro no kairyō*, 7.11 (November, 1925). Available online: http://library.jsce.or.jp/Image_DB/mag/dokai/pdf/07-11-0054.pdf, accessed May 25, 2015.

Miwa Shuzō 三輪修三. *Tamagawa* 多摩川, Yokohama: Yurindō, 1988.

Nagata Shirō 永田四郎. *Nihon semento sangyō-shi* 日本セメント産業史 [*A History of the Cement Industry in Japan*], Tokyo: Kensetsu Bunkasha, 1957.

Nish, Ian (ed.). *Britain and Japan: Biographical Portraits*, London: Routledge, 1997.

Oba, Osamu. Part Eight: "The Travels of Elephants," *Sino-Japanese Relations in the Edo Period*, 50–68. Available online: http://chinajapan.org/articles/11.2/11.2oba50-68.pdf, accessed May 25, 2015.

Ōta-ku Shi Hensan Iinkai (ed.). *Ōta-ku shi, shiryō-hen, minzoku* 大田区史·資料編·民俗 [*A History of Ota War, Documents on Folklore*], 1983.

Publications Committee, World Engineering Congress. *Industrial Japan: A Collection of Papers by Specialists on Various Branches of Industry in Japan*, Tokyo: International Publishing and Printing Co., 1929.

"Railways in Japan," Report by Consul-General Van Buren. Available online: http://books.google.co.jp/books?id=to08AQAAIAAJ&pg=PA688&dq=rokugo+river+bridge&hl=ja&sa=X&ei=TEj3UdzWNcHpkAXRi4DABw&ved=0CGIQ6AEwBzgo#v=onepage&q=rokugo%20river%20bridge&f=true, accessed May 25, 2015.

Salomon, M. A. "A Good Roads Movement in Japan: One Way in which American Efficiency Is Helping the Flowery Kingdom," *The Outlook*, March 26, 1919, pp. 12–13, available online: www.unz.org/Pub/Outlook-1919mar26-00515.

Seidensticker, E. G. *High City Low City*, Rutland, VT: Tuttle Publishers, 1984.

"Showa 2 nen rosei manga-shi 昭和二年道政漫画史 (A Cartoon History of Road Policy for 1927)," *Dōro no kairyō*, 10.1 (January, 1928). Available online: http://library.jsce.or.jp/Image_DB/mag/dokai/pdf/10-01-0188.pdf, accessed May 25, 2015.

Steele, M. William. "The History of the Tama River: Social Reconstructions," in T. Tvedt and E. Jakobsson (eds.), *A History of Water*, Vol. 1: Water Control and River Biographies, London: I. B. Tauris, 2006, pp. 217–238.

Steele, M. William. "The Making of a Bicycle Nation: Japan," *Transfers*, 2.2 (Summer 2012): 76–79.

"Taishō 15 nen manga-shi 大正十五年漫画史 (A Cartoon History for 1926)," *Dōro no kairyō*, 9.1 (January, 1927). Available online: http://library.jsce.or.jp/Image_DB/mag/dokai/pdf/09-01-0190.pdf, accessed May 25, 2015.

Takada Kei 高田景. "Keihin kokudō Rokugō-hashi to Tsurumi-hashi 京浜国道六郷橋と鶴見橋 (The Rokugō Bridge, the Connecting Link in the New Tokyo Yokohama Road, Now Nearing Completion)," *Doboku kenchiku kōji gahō* 土木建築工事画報 [*An Illustrated Construction Review*], 2.1 (January 1926): 31–34. Available online: http://library.jsce.or.jp/Image_DB/mag/gaho/kenchikukouji/02-01/02-01-0229.pdf, accessed May 25, 2015.

Takebe Kenichi 高部健一. *Michi no hanashi* 道の話 [*The Story of Roads*], Tokyo: Gihōdō Shuppan, 2000, Vol. 1, pp. 135–140.

Tamagawa-shi 多摩川誌 [*A History of the Tama River*], Tokyo: Kasen Kankyō Kanri Zaidan 河川環境管理財団, 1986; revised edition, 1994.

Tanaka Yoshimi 田中好. "Kairyō sareta Keihin kokudō 改良された京浜国道 (Improvements on the Keihin Highway)," *Dōro no kaiyō*, 9.1 (January 1927): 142–146.

"The New Railroad Bridge at Kawasaki," *Japan Weekly Mail*, October 6, 1877.

"The Other Costs of Cement," *Japan Times*, May 19, 2013. Available online: www.japantimes.co.jp/life/2013/05/19/environment/the-other-costs-of-concrete/#.Ugn4KR-b8ETM, accessed May 25, 2015.

Vaporis, Constantine. *Breaking Barriers: Travel and the State in Early Modern Japan*, Cambridge, MA: Harvard University Press, 1994.

United States Bureau of the Census (ed.). *Historical Statistics of the United States, Colonial Times to 1957*, U.S. Department of Commerce, 1960.

Yamada, Junji 山田順治. *Konkuriito no bunka-shi* コンクリートの文化史 [*A Cultural History of Concrete*], Tokyo: Bunichi Sōgō Shuppan, 1986.

Yamamoto Hirofumi (ed.). *Technological Innovation and the Development of Transportation in Japan*, Tokyo: United Nations University Press, 1993.

Index

Page numbers in *italics* denote tables, those in **bold** denote figures.

For Product Safety Concerns and Information please contact our EU representative GPSR@taylorandfrancis.com
Taylor & Francis Verlag GmbH, Kaufingerstraße 24, 80331 München, Germany

www.ingramcontent.com/pod-product-compliance
Ingram Content Group UK Ltd.
Pitfield, Milton Keynes, MK11 3LW, UK
UKHW021303190926
475701UK00009B/632

* 9 7 8 1 1 3 8 4 8 2 5 4 8 *